A Probability and
Statistics Companion

A Probability and Statistics Companion

John J. Kinney

WILEY

A JOHN WILEY & SONS, INC., PUBLICATION

Published by John Wiley & Sons, Inc., Hoboken, New Jersey
Published simultaneously in Canada.

For general information on our other products and services or for technical support, please contact our Customer Care Department within the United States at (800) 762-2974, outside the United States at (317) 572-3993 or fax (317) 572-4002.

Wiley also publishes its books in a variety of electronic formats. Some content that appears in print may not be available in electronic formats. For more information about Wiley products, visit our web site at www.wiley.com.

Library of Congress Cataloging-in-Publication Data:

Kinney, John J.
   A probability and statistics companion / John J. Kinney.
      p. cm.
   Includes bibliographical references and index.
   ISBN 978-0-470-47195-1 (pbk.)
   1. Probabilities. I. Title.

   QA273.K494 2009
   519.2–dc22

                                        2009009329

Typeset in 10/12pt Times by Thomson Digital, Noida, India.
Printed in the United States of America

10 9 8 7 6 5 4 3 2 1

For Cherry, Kaylyn, and James

Contents

Preface   **xv**

1. Probability and Sample Spaces   **1**

Why Study Probability?   1

Probability   2

Sample Spaces   2

Some Properties of Probabilities   8

Finding Probabilities of Events   11

Conclusions   16

Explorations   16

2. Permutations and Combinations: Choosing the Best Candidate;
Acceptance Sampling   **18**

Permutations   19
    Counting Principle   19
    Permutations with Some Objects Alike   20
    Permuting Only Some of the Objects   21

Combinations   22
    General Addition Theorem and Applications   25

Conclusions   35

Explorations   35

3. Conditional Probability   **37**

Introduction   37

Some Notation   40

Bayes' Theorem   45

Conclusions   46

Explorations   46

4. Geometric Probability   48

Conclusion   56

Explorations   57

5. Random Variables and Discrete Probability Distributions—Uniform, Binomial, Hypergeometric, and Geometric Distributions   58

Introduction   58

Discrete Uniform Distribution   59
    Mean and Variance of a Discrete Random Variable   60
    Intervals, σ, and German Tanks   61
    Sums   62

Binomial Probability Distribution   64
    Mean and Variance of the Binomial Distribution   68
    Sums   69

Hypergeometric Distribution   70
    Other Properties of the Hypergeometric Distribution   72

Geometric Probability Distribution   72

Conclusions   73

Explorations   74

6. Seven-Game Series in Sports   75

Introduction   75

Seven-Game Series   75

Winning the First Game   78
    How Long Should the Series Last?   79

Conclusions   81

Explorations   81

7. Waiting Time Problems   83

Waiting for the First Success   83

The Mythical Island   84

Waiting for the Second Success   85

Waiting for the rth Success   87

Mean of the Negative Binomial   87

Collecting Cereal Box Prizes   88

Heads Before Tails   88

Waiting for Patterns   90

Expected Waiting Time for HH   91

Expected Waiting Time for TH   93

An Unfair Game with a Fair Coin   94

Three Tosses   95

Who Pays for Lunch?   96

Expected Number of Lunches   98

Negative Hypergeometric Distribution   99

Mean and Variance of the Negative Hypergeometric   101

Negative Binomial Approximation   103

The Meaning of the Mean   104
    First Occurrences   104
    Waiting Time for c Special Items to Occur   104
    Estimating k   105

Conclusions   106

Explorations   106

**8. Continuous Probability Distributions: Sums, the Normal
   Distribution, and the Central Limit Theorem; Bivariate
   Random Variables**   **108**

Uniform Random Variable   109

Sums   111
    A Fact About Means   111

Normal Probability Distribution   113
    Facts About Normal Curves   114

Bivariate Random Variables   115
    Variance   119

Central Limit Theorem: Sums   121

Central Limit Theorem: Means   123

Central Limit Theorem   124

Expected Values and Bivariate Random Variables   124

   Means and Variances of Means   124

A Note on the Uniform Distribution   126

Conclusions   128

Explorations   129

9. Statistical Inference I   130

Estimation   131

Confidence Intervals   131

Hypothesis Testing   133

β and the Power of a Test   137

p-Value for a Test   139

Conclusions   140

Explorations   140

10. Statistical Inference II: Continuous Probability Distributions II—Comparing Two Samples   141

The Chi-Squared Distribution   141

Statistical Inference on the Variance   144

Student t Distribution   146

Testing the Ratio of Variances: The F Distribution   148

Tests on Means from Two Samples   150

Conclusions   154

Explorations   154

11. Statistical Process Control   155

Control Charts   155

Estimating σ Using the Sample Standard Deviations   157

Estimating σ Using the Sample Ranges   159

Control Charts for Attributes   161
    np Control Chart   161
    p Chart   163

Some Characteristics of Control Charts   164

Some Additional Tests for Control Charts   165

Conclusions   168

Explorations   168

12. Nonparametric Methods   170

Introduction   170

The Rank Sum Test   170

Order Statistics   173
    Median   174
    Maximum   176
    Runs   180
       Some Theory of Runs   182

Conclusions   186

Explorations   187

13. Least Squares, Medians, and the Indy 500   188

Introduction   188

Least Squares   191
    Principle of Least Squares   191

Influential Observations   193

The Indy 500   195

A Test for Linearity: The Analysis of Variance   197
    A Caution   201

Nonlinear Models   201

The Median–Median Line   202
    When Are the Lines Identical?   205
    Determining the Median–Median Line   207

Analysis for Years 1911–1969   209

Conclusions   210

Explorations   210

14. Sampling   211

Simple Random Sampling   212

Stratification   214
  Proportional Allocation   215
  Optimal Allocation   217

Some Practical Considerations   219

Strata   221

Conclusions   221

Explorations   221

15. Design of Experiments   223

Yates Algorithm   230

Randomization and Some Notation   231

Confounding   233

Multiple Observations   234

Design Models and Multiple Regression Models   235

Testing the Effects for Significance   235

Conclusions   238

Explorations   238

16. Recursions and Probability   240

Introduction   240

Conclusions   250

Explorations   250

17. Generating Functions and the Central Limit Theorem   **251**

Means and Variances   253

A Normal Approximation   254

Conclusions   255

Explorations   255

Bibliography   **257**

Where to Learn More   257

Index   **259**

Preface

Courses in probability and statistics are becoming very popular, both at the college and at the high school level, primarily because they are crucial in the analysis of data derived from samples and designed experiments and in statistical process control in manufacturing. Curiously, while these topics have put statistics at the forefront of scientific investigation, they are given very little emphasis in textbooks for these courses.

This book has been written to provide instructors with material on these important topics so that they may be included in introductory courses. In addition, it provides instructors with examples that go beyond those commonly used. I have developed these examples from my own long experience with students and with teachers in teacher enrichment programs. It is hoped that these examples will be of interest in themselves, thus increasing student motivation in the subjects and providing topics students can investigate in individual projects.

Although some of these examples can be regarded as advanced, they are presented here in ways to make them accessible at the introductory level. Examples include a problem involving a run of defeats in baseball, a method of selecting the best candidate from a group of applicants for a position, and an interesting set of problems involving the waiting time for an event to occur.

Connections with geometry are frequent. The fact that the medians of a triangle meet at a point becomes an extremely useful fact in the analysis of bivariate data; problems in conditional probability, often a challenge for students, are solved using only the area of a rectangle. Graphs allow us to see many solutions visually, and the computer makes graphic illustrations and heretofore exceedingly difficult computations quick and easy.

Students searching for topics to investigate will find many examples in this book.

I think then of the book as providing both supplemental applications and novel explanations of some significant topics, and trust it will prove a useful resource for both teachers and students.

It is a pleasure to acknowledge the many contributions of Susanne Steitz-Filler, my editor at John Wiley & Sons. I am most deeply grateful to my wife, Cherry; again, she has been indispensable.

John Kinney
Colorado Springs
April 2009

Chapter 1

Probability and Sample Spaces

CHAPTER OBJECTIVES:

- to introduce the theory of probability
- to introduce sample spaces
- to show connections with geometric series, including a way to add them without a formula
- to show a use of the Fibonacci sequence
- to use the binomial theorem
- to introduce the basic theorems of probability.

WHY STUDY PROBABILITY?

There are two reasons to study probability. One reason is that this branch of mathematics contains many interesting problems, some of which have very surprising solutions. Part of its fascination is that some problems that appear to be easy are, in fact, very difficult, whereas some problems that appear to be difficult are, in fact, easy to solve. We will show examples of each of these types of problems in this book. Some problems have very beautiful solutions.

The second, and compelling, reason to study probability is that it is the mathematical basis for the statistical analysis of experimental data and the analysis of sample survey data. Statistics, although relatively new in the history of mathematics, has become a central part of science. Statistics can tell experimenters what observations to take so that conclusions to be drawn from the data are as broad as possible. In sample surveys, statistics tells us how many observations to take (usually, and counterintuitively, relatively small samples) and what kinds of conclusions can be taken from the sample data.

A Probability and Statistics Companion, John J. Kinney
Copyright © 2009 by John Wiley & Sons, Inc.

Each of these areas of statistics is discussed in this book, but first we must establish the probabilistic basis for statistics.

Some of the examples at the beginning may appear to have little or no practical application, but these are needed to establish ideas since understanding problems involving actual data can be very challenging without doing some simple problems first.

PROBABILITY

A brief introduction to probability is given here with an emphasis on some unusual problems to consider for the classroom. We follow this chapter with chapters on permutations and combinations, conditional probability, geometric probability, and then with a chapter on random variables and probability distributions.

We begin with a framework for thinking about problems that involve randomness or chance.

SAMPLE SPACES

An experimenter has four doses of a drug under testing and four doses of an inert placebo. If the drugs are randomly allocated to eight patients, what is the probability that the experimental drug is given to the first four patients?

This problem appears to be very difficult. One of the reasons for this is that we lack a framework in which to think about the problem. Most students lack a structure for thinking about probability problems in general and so one must be created. We will see that the problem above is in reality not as difficult as one might presume.

Probability refers to the *relative frequency* with which events occur where there is some element or randomness or chance. We begin by enumerating, or showing, the set of all the possible outcomes when an experiment involving randomness is performed. This set is called a *sample space*.

We will not solve the problem involving the experimental drug here but instead will show other examples involving a sample space.

EXAMPLE 1.1   *A Production Line*

Items coming off a production line can be classified as either good (G) or defective (D). We observe the next item produced.

Here the set of all possible outcomes is

$$S = \{G, D\}$$

since one of these sample points must occur.

Now suppose we inspect the next five items that are produced. There are now 32 sample points that are shown in Table 1.1.

Table 1.1

Point	Good	Runs	Point	Good	Runs
GGGGG	5	1	GGDDD	2	2
GGGGD	4	2	GDGDD	2	4
GGGDG	4	3	DGGDD	2	3
GGDGG	4	3	DGDGD	2	5
GDGGG	4	3	DDGGD	2	3
DGGGG	4	2	DDGDG	2	4
DGGGD	3	3	DDDGG	2	2
DGGDG	3	4	GDDGD	2	4
DGDGG	3	4	GDDDG	2	3
DDGGG	3	2	GDDGD	2	4
GDDGG	3	3	GDDDD	1	2
GDGDG	3	5	DGDDD	1	3
GDGGD	3	4	DDGDD	1	3
GGDDG	3	3	DDDGD	1	3
GGDGD	3	4	DDDDG	1	2
GGGDD	3	2	DDDDD	0	1

We have shown in the second column the number of good items that occur with each sample point. If we collect these points together we find the distribution of the number of good items in Table 1.2.

It is interesting to see that these frequencies are exactly those that occur in the binomial expansion of

$$2^5 = (1 + 1)^5 = 1 + 5 + 10 + 10 + 5 + 1 = 32$$

This is not coincidental; we will explain this subsequently.

The sample space also shows the number of *runs* that occur. A *run* is a sequence of like adjacent results of length 1 or more, so the sample point $GGDGG$ contains three runs while the sample point $GDGDD$ has four runs.

It is also interesting to see, in Table 1.3, the frequencies with which various numbers of runs occur.

Table 1.2

Good	Frequency
0	1
1	5
2	10
3	10
4	5
5	1

Table 1.3

Runs	Frequency
1	2
2	8
3	12
4	8
5	2

We see a pattern but not one as simple as the binomial expansion we saw previously. So we see that like adjacent results are almost certain to occur somewhere in the sequence that is the sample point. The mean number of runs is 3. If a group is asked to write down a sequence of, say, G's and D's, they are likely to write down too many runs; like symbols are very likely to occur together. In a baseball season of 162 games, it is virtually certain that runs of several wins or losses will occur. These might be noted as remarkable in the press; they are not. We will explore the topic of runs more thoroughly in Chapter 12.

One usually has a number of choices for the sample space. In this example, we could choose the sample space that has 32 points or the sample space {0, 1, 2, 3, 4, 5} indicating the number of good items or the set {1, 2, 3, 4, 5} showing the number of runs. So we have three possible useful sample spaces.

Is there a "correct" sample space? The answer is "no". The sample space chosen for an experiment depends upon the *probabilities* one wishes to calculate. Very often one sample space will be much easier to deal with than another for a problem, so alternative sample spaces provide different ways for viewing the same problem. As we will see, the probabilities assigned to these sample points are quite different.

We should also note that good and defective items usually do not come off production lines at random. Items of the same sort are likely to occur together. The frequency of defective items is usually extremely small, so the sample points are by no means equally likely. We will return to this when we consider *acceptance sampling in* Chapter 2 and *statistical process control in* Chapter 11. ∎

EXAMPLE 1.2   *Random Arrangements*

The numbers 1, 2, 3, and 4 are arranged in a line at random.

The sample space here consists of all the possible orders, as shown below.

$$
S = \begin{cases}
1234^* & 2134^* & 3124^* & 4123 \\
1243^* & 2143 & 3142 & 4132^* \\
1324^* & 2314^* & 3214^* & 4231^* \\
1342^* & 2341 & 3241^* & 4213^* \\
1423^* & 2413 & 3412 & 4312 \\
1432^* & 2431^* & 3421 & 4321
\end{cases}
$$

S here contains 24 elements, the number of possible linear orders, or arrangements of 4 distinct items. These arrangements are called *permutations*. We will consider permutations more generally in Chapter 2.

A well-known probability problem arises from the above permutations. Suppose the "natural" order of the four integers is 1234. If the four integers are arranged randomly, how many of the integers occupy their own place? For example, in the order 3214, the integers 2 and 4 are in their own place. By examining the sample space above, it is easy to count the permutations in which at least one of the integers is in its own place. These are marked with an asterisk in *S*. We find 15 such permutations, so $15/24 = 0.625$ of the permutations has at least one integer in its own place.

Now what happens as we increase the number of integers? This leads to the well-known "hat check" problem that involves *n* people who visit a restaurant and each check a hat, receiving a numbered receipt. Upon leaving, however, the hats are distributed at random. So the hats are distributed according to a random permutation of the integers $1, 2, \ldots, n$. What proportion of the diners gets his own hat?

If there are four diners, we see that 62.5% of the diners receive their own hats. Increasing the number of diners complicates the problem greatly if one is thinking of listing all the orders and counting the appropriate orders as we have done here. It is possible, however, to find the answer without proceeding in this way. We will show this in Chapter 2.

It is perhaps surprising, and counterintuitive, to learn that the proportion for 100 people differs little from that for 4 people! In fact, the proportion approaches $1 - 1/e = 0.632121$ as *n* increases. (To six decimal places, this is the exact result for 10 diners.) This is our first, but by no means our last, encounter with $e = 2.71828 \ldots$, the base of the system of natural logarithms. The occurrence of *e* in probability, however, has little to do with natural logarithms.   ■

The next example also involves *e*.

EXAMPLE 1.3   *Running Sums*

A box contains slips of paper numbered 1, 2, and 3, respectively. Slips are drawn one at a time, replaced, and a cumulative running sum is kept until the sum equals or exceeds 4.

This is an example of a *waiting time* problem; we wait until an event occurs. The event can occur in two, three, or four drawings. (It must occur no later than the fourth drawing.)

The sample space is shown in Table 1.4, where *n* is the number of drawings and the sample points show the order in which the integers were selected.

Table 1.4

n	Orders
2	(1,3),(3,1),(2,2) (2,3),(3,2),(3,3)
3	(1,1,2),(1,1,3),(1,2,1),(1,2,2) (1,2,3),(2,1,1),(2,1,2),(2,1,3)
4	(1,1,1,1),(1,1,1,2),(1,1,1,3)

Table 1.5

n	Expected value
1	2.00
2	2.25
3	2.37
4	2.44
5	2.49

We will show later that the *expected number* of drawings is 2.37.

What happens as the number of slips of paper increases? The approach used here becomes increasingly difficult. Table 1.5 shows exact results for small values of n, where we draw until the sum equals or exeeds $n + 1$.

While the value of n increases, the expected length of the game increases, but at a decreasing rate. It is too difficult to show here, but the expected length of the game approaches $e = 2.71828\ldots$ as n increases.

This does, however, make a very interesting classroom exercise either by generating random numbers within the specified range or by a computer simulation. The result will probably surprise students of calculus and be an interesting introduction to e for other students. ∎

EXAMPLE 1.4 *An Infinite Sample Space*

Examples 1.1, 1.2, and 1.3 are examples of finite sample spaces, since they contain a finite number of elements. We now consider an infinite sample space.

We observe a production line until a defective (D) item appears. The sample space now is infinite since the event may never occur. The sample space is shown below (where G denotes a good item).

$$S = \left\{ \begin{array}{c} D \\ GD \\ GGD \\ GGGD \\ \cdot \\ \cdot \\ \cdot \end{array} \right\}$$

We note that S in this case is a countable set, that is, a set that can be put in one-to-one correspondence with the set of positive integers. Countable sample spaces often behave as if they were finite. Uncountable infinite sample spaces are also encountered in probability, but we will not consider these here. ∎

EXAMPLE 1.5   *Tossing a Coin*

We toss a coin five times and record the tosses in order. Since there are two possibilities on each toss, there are $2^5 = 32$ sample points. A sample space is shown below.

$$S = \begin{cases} TTTTT & TTTTH & TTTHT & TTHTT & THTTT & HTTTT \\ HHTTT & HTHTT & HTTHT & TTTHH & THTHT & TTHHT \\ HTTHH & HTTTH & THTTH & TTHTH & HHHHT & THTHH \\ THHTH & THHHT & TTHHH & HTHTH & THHTT & HHHTT \\ HTHHT & HHTHT & HHHHT & HHHTH & HHTHH & HTHHH \\ THHHH & HHHHH \end{cases}$$

It is also possible in this example simply to count the number of heads, say, that occur. In that case, the sample space is

$$S_1 = \{0, 1, 2, 3, 4, 5\}$$

Both S and S_1 are sets that contain all the possibilities when the experiment is performed and so are sample spaces. So we see that the sample space is not uniquely defined. Perhaps one can think of other sets that describe the sample space in this case.   ∎

EXAMPLE 1.6   *AP Statistics*

A class in advanced placement statistics consists of three juniors (J) and four seniors (S). It is desired to select a committee of size two. An appropriate sample space is

$$S = \{JJ, JS, SJ, SS\}$$

where we have shown the class of the students selected in order. One might also simply count the number of juniors on the committee and use the sample space

$$S_1 = \{0, 1, 2\}$$

Alternatively, one might consider the individual students selected so that the sample space, shown below, becomes

$$S_2 = \{J_1 J_2, J_1 J_3, J_2 J_3, S_1 S_2, S_1 S_3, S_1 S_4, S_2 S_3, S_2 S_4, S_3 S_4,$$
$$J_1 S_1, J_1 S_2, J_1 S_3, J_1 S_4, J_2 S_1, J_2 S_2, J_2 S_3, J_2 S_4, J_3 S_1,$$
$$J_3 S_2, J_3 S_3, J_3 S_4\}$$

S_2 is as detailed a sample space one can think of, if order of selection is disregarded, so one might think that these 21 sample points are equally likely to occur provided no priority is given to any of the particular individuals. So we would expect that each of the points in S_2 would occur about $1/21$ of the time. We will return to assigning probabilities to the sample points in S and S_2 later in this chapter.   ∎

EXAMPLE 1.7   *Let's Make a Deal*

On the television program Let's Make a Deal, a contestant is shown three doors, only one of which hides a valuable prize. The contestant chooses one of the doors and the host then opens one of the remaining doors to show that it is empty. The host then asks the contestant if she wishes to change her choice of doors from the one she selected to the remaining door.

Let W denote a door with the prize and E_1 and E_2 the empty doors. Supposing that the contestant switches choices of doors (which, as we will see in a later chapter, she should do), and we write the contestant's initial choice and then the door she finally ends up with, the sample space is

$$S = \{(W, E_1), (W, E_2), (E_1, W), (E_2, W)\}$$

■

EXAMPLE 1.8   *A Birthday Problem*

A class in calculus has 10 students. We are interested in whether or not at least two of the students share the same birthday. Here the sample space, showing all possible birthdays, might consist of components with 10 items each. We can only show part of the sample space since it contains $365^{10} = 4.1969 \times 10^{25}$ points! Here

$$S = \{(\text{March 10, June 15, April 24, } \ldots), (\text{May 5, August 2, September 9, } \ldots)\}$$

It may seem counterintuitive, but we can calculate the probability that at least two of the students share the same birthday without enumerating all the points in S. We will return to this problem later.

■

Now we continue to develop the theory of probability.

SOME PROPERTIES OF PROBABILITIES

Any subset of a sample space is called an *event*. In Example 1.1, the occurrence of a good item is an event. In Example 1.2, the sample point where the number 3 is to the left of the number 2 is an event. In Example 1.3, the sample point where the first defective item occurs in an even number of items is an event. In Example 1.4, the sample point where exactly four heads occur is an event.

We wish to calculate the *relative likelihood*, or *probability*, of these events. If we try an experiment n times and an event occurs t times, then the relative likelihood of the event is t/n. We see that relative likelihoods, or probabilities, are numbers between 0 and 1. If A is an event in a sample space, we write $P(A)$ to denote the probability of the event A.

Probabilities are governed by these three axioms:

1. $P(S) = 1$.

2. $0 \le P(A) \le 1$.

3. If events A and B are disjoint, so that $A \cap B = \varnothing$, then $P(A \cup B) = P(A) + P(B)$.

Axioms 1 and 2 are fairly obvious; the probability assigned to the entire sample space must be 1 since by definition of the sample space some point in the sample space must occur and the probability of an event must be between 0 and 1. Now if an event A occurs with probability $P(A)$ and an event B occurs with probability $P(B)$ and if the events cannot occur together, then the relative frequency with which one or the other occurs is $P(A) + P(B)$. For example, if a prospective student decides to attend University A with probability 2/5 and to attend University B with probability 1/5, she will attend one or the other (but not both) with probability $2/5 + 1/5 = 3/5$. This explains Axiom 3.

It is also very useful to consider an event, say A, as being composed of distinct points, say a_i, with probabilities $p(a_i)$. By Axiom 3 we can add these individual probabilities to compute $P(A)$ so

$$P(A) = \sum_{i=1}^{n} p(a_i)$$

It is perhaps easiest to consider a finite sample space, but our conclusions also apply to a countably infinite sample space. Example 1.4 involved a countable infinite sample space; we will encounter several more examples of these sample spaces in Chapter 7.

Disjoint events are also called *mutually exclusive* events.

Let $\overline{A}$ denote the points in the sample space where event A does not occur. Note that A and $\overline{A}$ are mutually exclusive so

$$P(S) = P(A \cup \overline{A}) = P(A) + P(\overline{A}) = 1$$

and so we have

Fact 1. $P(\overline{A}) = 1 - P(A)$.
Axiom 3 concerns events that are mutually exclusive. What if they are not mutually exclusive?

Refer to Figure 1.1.

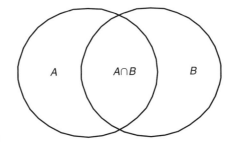

Figure 1.1

If we find $P(A) + P(B)$ by adding the probabilities of the distinct points in those events, then we have counted $P(A \cap B)$ twice so

Fact 2. $PA \cup B) = P(A) + P(B) - P(A \cap B)$,
where Fact 2 applies whether events A and B are disjoint or not.

Fact 2 is known as the addition theorem for two events. It can be generalized to three or more events in Fact 3:

Fact 3. (*General addition theorem*).

$$P(A_1 \cup A_2 \cup \cdots \cup A_n) = \sum_{i=n}^{n} P(A_i) - \sum_{i \neq j=1}^{n} P(A_i \cap A_j)$$

$$+ \sum_{i \neq j \neq k=1}^{n} P(A_i \cap A_j \cap A_k) - \cdots$$

$$\pm \sum_{i \neq j \neq \cdots \neq n=1}^{n} P(A_i \cap A_j \cap \cdots \cap A_n)$$

We simply state this theorem here. We prefer to prove it using techniques developed in Chapter 2, so we delay the proof until then.

Now we turn to events that can occur together.

EXAMPLE 1.9   *Drawing Marbles*

Suppose we have a jar of marbles containing five red and seven green marbles. We draw them out one at a time (without replacing the drawn marble) and want the probability that the first drawn marble is red and the second green. Clearly, the probability the first is red is 5/12. As for the second marble, the contents of the jar have now changed, and the probability the second marble is green given that the first marble is red is now 7/11. We conclude that the probability the first marble is red and the second green is $5/12 \cdot 7/11 = 35/132$. The fact that the composition of the jar changes with the selection of the first marble alters the probability of the color of the second marble.

The probability the second marble is green given that the first is red, 7/11, differs from the probability the marble is green, 7/12.

We call the probability the second marble is green, given the first is red, the *conditional probability* of a green, given a red. We say in general that

$$P(A \cap B) = P(A) \cdot P(B|A)$$

where we read $P(B|A)$ as the *conditional probability* of event B *given that event A has occurred.* This is called the *multiplication rule.*

Had the first marble been replaced before making the second drawing, the probability of drawing a green marble on the second drawing would have been the same as drawing a green marble on the first drawing, 7/12. In this case, $P(B|A) = P(B)$, and the events are called *independent*.   ∎

We will study conditional probability in Chapter 3.

Independent events and disjoint events are commonly confused. Independent events refer to events that can occur together; disjoint events cannot occur together. We refer now to events that do not have probability 0 (such events are encountered in nondenumerably infinite sample spaces).

If events are independent, then they cannot be disjoint since they must be able to occur together; if events are disjoint, then they cannot be independent because they cannot occur together.

FINDING PROBABILITIES OF EVENTS

The facts about probabilities, as shown in the previous section, are fairly easy. The difficulty arises when we try to apply them.

The first step in any probability problem is to define an appropriate sample space. More than one sample space is possible; it is usually the case that if order is considered, then the desired probabilities can be found, because that is the most detailed sample space one can write, but it is not always necessary to consider order.

Let us consider the examples for which we previously found sample spaces.

EXAMPLE 1.10   *A Binomial Problem*

In Example 1.1, we examined an item emerging from a production line and observed the result. It might be sensible to assign the probabilities to the events as $P(G) = 1/2$ and $P(D) = 1/2$ if we suppose that the production line is not a very good one. This is an example of a *binomial* event (where one of the two possible outcomes occurs at each trial) but it is not necessary to assign equal probabilities to the two outcomes.

It is a common error to presume, because a sample space has n points, that each point has probability $1/n$. For another example, when a student takes a course she will either pass it or fail it, but it would not be usual to assign equal probabilities to the two events. But if we toss a fair coin, then we might have $P(\text{Head}) = 1/2$ and $P(\text{Tail}) = 1/2$. We might also consider a loaded coin where $P(H) = p$ and $P(T) = q = 1 - p$ where, of course, $0 \le p \le 1$.

It is far more sensible to suppose in our production line example that $P(G) = 0.99$ and $P(D) = 0.01$ and even these assumptions assume a fairly poor production line. In that case, and assuming the events are independent, we then find that

$$P(GGDDG) = P(G) \cdot P(G) \cdot P(D) \cdot P(D) \cdot P(G) = (0.99)^3 \cdot (0.01)^2$$

$$= 0.000097$$

Also, since the sample points are disjoint, we can compute the probability we see exactly two defective items as

$$P(GGDDG)+P(DGDGG)+P(DGGDG)+P(DGGGD)+P(GDGGD)+P(GGDGD)$$

$$+P(GGGDD) + P(GDGDG) + P(GDDGG) + P(DGDGG) = 0.00097$$

Note that the probability above must be $10 \cdot P(GGDDG) = 10 \cdot (0.99)^3 \cdot (0.01)^2$ since each of the 10 orders shown above has the same probability. Note also that $10 \cdot (0.99)^3 \cdot (0.01)^2$ is a term in the binomial expansion $(0.99 + 0.01)^{10}$. ∎

We will consider more problems involving the binomial theorem in Chapter 5.

EXAMPLE 1.11   *More on Arrangements*

In Example 1.2, we considered all the possible permutations of four objects. Thinking that these permutations occur at random, we assign probability $1/24$ to each of the sample points. The events "3 occurs to the left of 2" then consists of the points

{3124, 3142, 4132, 1324, 3214, 1342, 3241, 3412, 4312, 1432, 3421, 4321}

Since there are 12 of these and since they are mutually disjoint and since each has probability $1/24$, we find

$$P(3 \text{ occurs to the left of } 2) = 12/24 = 1/2$$

We might have seen this without so much work if we considered the fact that in a random permutation, 3 is as likely to be to the left of 2 as to its right. As you were previously warned, easy looking problems are often difficult while difficult looking problems are often easy. It is all in the way one considers the problem. ∎

EXAMPLE 1.12   *Using a Geometric Series*

Example 1.4 is an example of a waiting time problem; that is, we do not have a determined number of trials, but we wait for an event to occur. If we consider the manufacturing process to be fairly poor and the items emerging from the production line are independent, then one possible assignment of probabilities is shown in Table 1.6.

Table 1.6

Event	Probability
D	0.01
GD	$0.99 \cdot 0.01 = 0.0099$
GGD	$(0.99)^2 \cdot 0.01 = 0.009801$
$GGGD$	$(0.99)^3 \cdot 0.01 = 0.009703$
$\vdots$	$\vdots$

We should check that the probabilities add up to 1. We find that (using S for sum now)

$$S = 0.01 + 0.99 \cdot 0.01 + (0.99)^2 \cdot 0.01 + (0.99)^3 \cdot 0.01 + \cdots$$

and so

$$0.99S = 0.99 \cdot 0.01 + (0.99)^2 \cdot 0.01 + (0.99)^3 \cdot 0.01 + \cdots$$

and subtracting one series from another we find

$$S - 0.99S = 0.01$$

or

$$0.01S = 0.01$$

and so $S = 1$.

This is also a good opportunity to use the geometric series to find the sum, but we will have to use for the above trick in later chapters for series that are not geometric.

What happens if we assign arbitrary probabilities to defective items and good items? This would certainly be the case with an effective production process. If we let $P(D) = p$ and $P(G) = 1 - p = q$, then the probabilities appear as shown in Table 1.7.

Table 1.7

Event	Probability
D	p
GD	qp
GGD	$q^2 p$
$GGGD$	$q^3 p$
$\vdots$	$\vdots$

Again, have we assigned a probability of 1 to the entire sample space? Letting S stand for sum again, we have

$$S = p + qp + q^2 p + q^3 p + \cdots$$

and so

$$qS = qp + q^2 p + q^3 p + \cdots$$

and subtracting, we find

$$S - qS = p$$

so

$$(1 - q)S = p$$

or

$$pS = p$$

meaning that $S = 1$.

This means that our assignment of probabilities is correct for any value of p.

Now let us find the probability the first defective item occurs at an even-numbered toss. Let the event be denoted by E.

$$P(E) = qp + q^3 p + q^5 p + \cdots$$

and so

$$q^2 \cdot P(E) = q^3 p + q^5 p + \cdots$$

and subtracting we find

$$P(E) - q^2 \cdot P(E) = qp$$

from which it follows that

$$(1 - q^2) \cdot P(E) = qp$$

and so

$$P(E) = \frac{qp}{1 - q^2} = \frac{qp}{(1-q)(1+q)} = \frac{q}{1+q}$$

If the process produces items with the above probabilities, this becomes $0.01/(1 + 0.01) = 0.0099$. One might presume that the probability the first defective item occurs at an even-numbered observation is the same as the probability the first defective item occurs at an odd-numbered observation. This cannot be correct, however, since the probability the first defective item occurs at the first observation (an odd-numbered observation) is p. It is easy to show that the probability the first defective item occurs at an odd-numbered observation is $1/(1 + q)$, and for a process with equal probabilities, such as tossing a fair coin, this is $2/3$.   ∎

EXAMPLE 1.13   *Relating Two Sample Spaces*

Example 1.5 considers a binomial event where we toss a coin five times. In the first sample space, S, we wrote out all the possible orders in which the tosses could occur. This is of course impossible if we tossed the coin, say, 10, 000 times! In the second sample space, S_1, we simply looked at the number of heads that occurred. The difference is that the sample points are not equally likely.

In the first sample space, where we enumerated the result of each toss, using the fact that the tosses are independent, and assuming that the coin is loaded, where $P(H) = p$ and $P(T) = 1 - p = q$, we find, to use two examples, that

$$P(TTTTT) = q^5 \quad \text{and} \quad P(HTTHH) = p^3 q^2$$

Now we can relate the two sample spaces. In S_1, $P(0) = P(0 \text{ heads}) = P(TTTTT) = q^5$.

Now $P(1 \text{ head})$ is more complex since the single head can occur in one of the five possible places. Since these sample points are mutually disjoint, $P(1 \text{ head}) = 5 \cdot p \cdot q^4$.

There are 10 points in S where two heads appear. Each of these points has probability $p^2 \cdot q^3$ so $P(2 \text{ heads}) = 10 \cdot p^2 \cdot q^3$.

We find, similarly, that $P(3 \text{ heads}) = 10 \cdot p^3 \cdot q^2$, $P(4 \text{ heads}) = 5 \cdot p^4 \cdot q$, and, finally, $P(5 \text{ heads}) = p^5$. So the sample points in S_1 are far from being equally likely. If we add all these probabilities, we find

$$q^5 + 5 \cdot p \cdot q^4 + 10 \cdot p^2 \cdot q^3 + 10 \cdot p^3 \cdot q^2 + 5 \cdot p^4 \cdot q + p^5$$

which we recognize as the binomial expansion $(q + p)^5$ that is 1 since $q = 1 - p$.

In a binomial situation (where one of the two possible outcomes occurs at each trial) with n observations, we see that the probabilities are the individual terms in the binomial expansion $(q + p)^n$. ∎

EXAMPLE 1.14   *Committees and Probability*

In Example 1.6, we chose a committee of two students from a class with three juniors and four seniors. The sample space we used is

$$S = \{JJ, JS, SJ, SS\}$$

How should probabilities be assigned to the sample points? First we realize that each sample point refers to a combination of events so that JJ means choosing a junior first *and* then choosing another junior. So JJ really refers to $J \cap J$ whose probability is

$$P(J \cap J) = P(J) \cdot P(J|J)$$

by the multiplication rule. Now $P(J) = 3/7$ since there are three juniors and we regard the selection of the students as equally likely. Now, with one of the juniors selected, we have only two juniors to choose from, so $P(J|J) = 2/6$ and so

$$P(J \cap J) = \frac{3}{7} \cdot \frac{2}{6} = \frac{1}{7}$$

In a similar way, we find

$$P(J \cap S) = \frac{3}{7} \cdot \frac{4}{6} = \frac{2}{7}$$

$$P(S \cap J) = \frac{4}{7} \cdot \frac{3}{6} = \frac{2}{7}$$

$$P(S \cap S) = \frac{4}{7} \cdot \frac{3}{6} = \frac{2}{7}$$

These probabilities add up to 1 as they should. ∎

EXAMPLE 1.15   *Let's Make a Deal*

Example 1.7 is the Let's Make a Deal problem. It has been widely written about since it is easy to misunderstand the problem. The contestant chooses one of the doors that we have labeled W, E_1, and E_2. We suppose again that the contestant switches doors after the host exhibits one of the nonchosen doors to be empty.

If the contestant chooses W, then the host has two choices of empty doors to exhibit. Suppose he chooses these with equal probabilities. Then $W \cap E_1$ means that the contestant initially chooses W, the host exhibits E_2, the contestant switches doors and ends up with E_1. The probability of this is then

$$P(W \cap E_1) = P(W) \cdot P(E_1|W) = \frac{1}{3} \cdot \frac{1}{2} = \frac{1}{6}$$

In an entirely similar way,

$$P(W \cap E_2) = P(W) \cdot P(E_2|W) = \frac{1}{3} \cdot \frac{1}{2} = \frac{1}{6}$$

Using the switching strategy, the only way the contestant loses is by selecting the winning door first (and then switching to an empty door), so the probability the contestant loses is the sum of these probabilities, $1/6 + 1/6 = 1/3$, which is just the probability of choosing W in the first place. It follows that the probability of winning under this strategy is 2/3!

Another way to see this is to calculate the probabilities of the two ways to winning, namely, $P(E_1 \cap W)$ and $P(E_2 \cap W)$. In either of these, an empty door is chosen first. This means that the host has only one choice for exhibiting an empty door. So each of these probabilities is simply the probability of choosing the specified empty door first, which is 1/3. The sum of these probabilities is 2/3, as we found before.

After the contestant selects a door, the probability the winning door is one not chosen is 2/3. The fact that one of these is shown to be empty does not change this probability.   ■

EXAMPLE 1.16   *A Birthday Problem*

To think about the birthday problem, in Example 1.8, we will use the fact that $P(\overline{A}) = 1 - P(A)$. So if A denotes the event that the birthdays are all distinct, then $\overline{A}$ denotes the event that at least two of the birthdays are the same.

To find $P(A)$, note that the first person can have any birthday in the 365 possible birthdays, the next can choose any day of the 364 remaining, the next has 363 choices, and so on.

If there are 10 students in the class, then

$$P(\text{at least two birthdays are the same}) = 1 - \frac{365}{365} \cdot \frac{364}{365} \cdot \frac{363}{365} \cdot \ldots \cdot \frac{356}{365} = 0.116948$$

If there are n students, we find

$$P(\text{at least two birthdays are the same}) = 1 - \frac{365}{365} \cdot \frac{364}{365} \cdot \frac{363}{365} \cdot \ldots \cdot \frac{366 - (n-1)}{365}$$

This probability increases as n increases. It is slightly more than $1/2$ if $n = 23$, while if $n = 40$, it is over 0.89. These calculations can be made with your graphing calculator.

This result may be surprising, but note that *any* two people in the group can share a birthday; this is not the same as finding someone whose birthday matches, say, your birthday.   ■

CONCLUSIONS

This chapter has introduced the idea of the probability of an event and has given us a framework, called a sample space, in which to consider probability problems. The axioms on which probability is based have been shown and some theorems resulting from them have been shown.

EXPLORATIONS

1. Consider all the arrangements of the integers 1, 2, 3, and 4. Count the number of *derangements*, that is, the number of arrangements in which no integer occupies its own place. Speculate on the relative frequency of the number of derangements as the number of integers increases.

2. Simulate the *Let's Make a Deal* problem by taking repeated selections from three cards, one of which is designated to be the prize. Compare two strategies: (1) never changing the selection and (2) always changing the selection.

3. A hat contains tags numbered 1, 2, 3, 4, 5, 6. Two tags are selected. Show the sample space and then compare the probability that the number on the second tag exceeds the number on the first tag when (a) the first tag is not replaced before the second tag is drawn and (b) the first tag is replaced before the second tag is drawn.

4. Find the probability of (a) exactly three heads and (b) at most three heads when a fair coin is tossed five times.

5. If p is the probability of obtaining a 5 at least once in n tosses of a fair die, what is the least value of n so that $p \geq 1/2$?

6. Simulate drawing integers from the set 1, 2, 3 until the sum exceeds 4. Compare your mean value to the expected value given in the text.

7. Toss a fair coin 100 times and find the frequencies of the number of runs. Repeat the experiment as often as you can.

8. Use a computer to simulate tossing a coin 1000 times and find the frequencies of the number of runs produced.

Chapter 2

Permutations and Combinations: Choosing the Best Candidate; Acceptance Sampling

CHAPTER OBJECTIVES:

- to discuss permutations and combinations
- to use the binomial theorem
- to show how to select the best candidate for a position
- to encounter an interesting occurrence of *e*
- to show how sampling can improve the quality of a manufactured product
- to use the principle of maximum likelihood
- to apply permutations and combinations to other practical problems.

An executive in a company has an opening for an executive assistant. Twenty candidates have applied for the position. The executive is constrained by company rules that say that candidates must be told whether they are selected or not at the time of an interview. How should the executive proceed so as to maximize the chance that the best candidate is selected?

Manufacturers of products commonly submit their product to inspection before the product is shipped to a consumer. This inspection usually measures whether or not the product meets the manufacturer's as well as the consumer's specifications. If the product inspection is destructive, however (such as determining the length of time a light bulb will burn), then all the manufactured product cannot be inspected.

A Probability and Statistics Companion, John J. Kinney
Copyright © 2009 by John Wiley & Sons, Inc.

Even if the inspection is not destructive or harmful to the product, inspection of all the product manufactured is expensive and time consuming. If the testing is destructive, it is possible to inspect only a random sample of the product produced. Can random sampling improve the quality of the product sold?

We will consider each of these problems, as well as several others, in this chapter. First we must learn to count points in sets, so we discuss *permutations* and *combinations* as well as some problems solved using them.

PERMUTATIONS

A *permutation* is a linear arrangement of objects, or an arrangement of objects in a row, in which the *order* of the objects is important.

For example, if we have four objects, which we will denote by a, b, c, and d, there are 24 distinct linear arrangements as shown in Table 2.1

Table 2.1

abcd	abdc	acbd	acdb	adbc	adcb
bacd	badc	bcda	bcad	bdac	bdca
cabd	cadb	cbda	cbad	cdab	cdba
dabc	dacb	dbca	dbac	dcab	dcba

In Chapter 1, we showed all the permutations of the set {1, 2, 3, 4} and, of course, found 24 of these. To count the permutations, we need a fundamental principle first.

Counting Principle

Fundamental Counting Principle. If an event A can occur in n ways and an event B can occur in m ways, then A and B can occur in $n \cdot m$ ways.

The proof of this can be seen in Figure 2.1, where we have taken $n = 2$ and $m = 3$. There are $2 \cdot 3 = 6$ paths from the starting point. It is easy to see that the branches of the diagram can be generalized.

The counting principle can be extended to three or more events by simply multiplying the number of ways subsequent events can occur. The number of permutations of the four objects shown above can be counted as follows, assuming for the moment that all of the objects to be permuted are distinct.

Since there are four positions to be filled to determine a unique permutation, we have four choices for the letter or object in the leftmost position. Proceeding to the right, there are three choices of letter or object to be placed in the next position. This gives $4 \cdot 3$ possible choices in total. Now we are left with two choices for the object in the next position and finally with only one choice for the rightmost position. So there are $4 \cdot 3 \cdot 2 \cdot 1 = 24$ possible permutations of these four objects. We have made repeated use of the counting principle.

We denote $4 \cdot 3 \cdot 2 \cdot 1$ as 4! (read as "4 factorial").

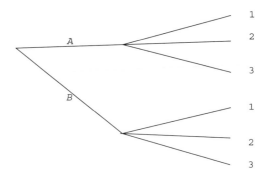

Figure 2.1

It follows that there are $n!$ possible permutations of n distinct objects.

The number of permutations of n distinct objects grows rapidly as n increases as shown in Table 2.2.

Table 2.2

n	$n!$
1	1
2	2
3	6
4	24
5	120
6	720
7	5,040
8	40,320
9	362,880
10	3,628,800

Permutations with Some Objects Alike

Sometimes not all of the objects to be permuted are distinct. For example, suppose we have 3, A's, 4 B's, and 5 C's to be permuted, or 12 objects all together. There are not 12! permutations, since the A's are not distinguishable from each other, nor are the B's, nor are the C's.

Suppose we let G be the number of distinct permutations and that we have a list of these permutations. Now number the A's from 1 to 3. These can be permuted in 3! ways; so, if we permute the A's in each item in our list, the list now has $3!G$ items. Now label the B's from 1 to 4 and permute the B's in each item in the list in all 4! ways. The list now has $4!3!G$ items. Finally, number the 5 C's and permute these for each item in the list. The list now contains $5!4!3!G$ items. But now each of the items is distinct, so the list has 12! items. We see that $5!4!3!G = 12!$, so $G = \frac{12!}{5!4!3!}$ or $G = 27,720$ and this is considerably less than $12! = 479,001,600$.

Permuting Only Some of the Objects

Now suppose that we have n distinct objects and we wish to permute r of them, where $r \leq n$. We now have r boxes to fill. This can be done in

$$n \cdot (n-1) \cdot (n-2) \cdots [n-(r-1)] = n \cdot (n-1) \cdot (n-2) \cdots (n-r+1)$$

ways. If $r < n$, this expression is not a factorial, but can be expressed in terms of factorials by multiplying and dividing by $(n-r)!$ We see that

$$
\begin{aligned}
&\quad n \cdot (n-1) \cdot (n-2) \cdots (n-r+1) \\
&= \frac{n \cdot (n-1) \cdot (n-2) \cdots (n-r+1) \cdot (n-r)!}{(n-r)!} \\
&= \frac{n!}{(n-r)!}
\end{aligned}
$$

We will have little use for this formula. We derived it so that we can count the number of samples that can be chosen from a population, which we do subsequently. For the formula to work for any value of r, we define $0! = 1$.

We remark now that the 20 applicants to the executive faced with choosing a new assistant could appear in $20! = 24,329,020,081,766,400,000$ different orders. Selecting the best of the group by making a random choice means that the best applicant has a $1/20 = 0.05$ chance of being selected, a fairly low probability. So the executive must create a better procedure. The executive can, as we will see, choose the best candidate with a probability approaching $1/3$, but that is something we will discuss much later.

There are 52! distinct arrangements of a deck of cards. This number is of the order $8 \cdot 10^{67}$. It is surprising to find, if we could produce 10,000 distinct permutations of these *per second,* that it would take about $2 \cdot 10^{56}$ years to enumerate all of these. We usually associate impossible events with infinite sets, but this is an example of a finite set for which this event is impossible.

For example, suppose we have four objects ($a, b, c,$ and d again) and that we wish to permute only two of these. We have four choices for the leftmost position and three choices for the second position, giving $4 \cdot 3 = 12$ permutations.

Applying the formula we have $n = 4$ and $r = 2$, so

$$_4P_2 = \frac{4!}{(4-2)!} = \frac{4!}{2!} = \frac{4 \cdot 3 \cdot 2!}{2!} = 4 \cdot 3 = 12$$

giving the correct result.

Permutations are often the basis for a sample space in a probability problem. Here are two examples.

EXAMPLE 2.1   *Lining Up at a Counter*

Jim, Sue, Bill, and Kate stand in line at a ticket counter. Assume that all the possible permutations, or orders, are equally likely. There are $4! = 24$ of these permutations. If we want to find the probability that Sue is in the second place, we must count the number of ways in which she could be in the second place. To count these, first put her there—there is only one way to do this. This leaves three choices for the first place, two choices for the third place, and, finally, only one choice for the fourth place. There are then $3 \cdot 1 \cdot 2 \cdot 1 = 6$ ways for Sue to occupy the second place. So

$$P(\text{Sue is in the second place}) = \frac{3 \cdot 1 \cdot 2 \cdot 1}{4 \cdot 3 \cdot 2 \cdot 1} = \frac{6}{24} = \frac{1}{4}$$

This is certainly no surprise. We would expect that any of the four people has a probability of $1/4$ to be in any of the four positions. ■

EXAMPLE 2.2   *Arranging Marbles*

Five red and seven blue marbles are arranged in a row. We want to find the probability that both the end marbles are red.

Number the marbles from 1 to 12, letting the red marbles be numbered from 1 to 5 for convenience. The sample space consists of all the possible permutations of 12 distinct objects, so the sample space contains $12!$ points, each of which, we will assume, is equally likely. Now we must count the number of points in which the end points are both red. We have five choices for the marble at the left end and four choices for the marble at the right end. The remaining marbles, occupying places between the ends, can be arranged in $10!$ ways, so

$$P(\text{end marbles are both red}) = \frac{5 \cdot 4 \cdot 10!}{12!} = \frac{5 \cdot 4 \cdot 10!}{12 \cdot 11 \cdot 10!} = \frac{5 \cdot 4}{12 \cdot 11} = \frac{5}{33}$$

■

COMBINATIONS

If we have n distinct objects and we choose only r of them, we denote the number of possible samples, *where the order in which the sample items are selected is of no importance*, by $\binom{n}{r}$, which we read as "n choose r". We want to find a formula for this quantity and first we consider a special case. Return to the problem of counting the number of samples of size 3 that can be chosen from the set $\{1, 2, 3, 4, 5, 6, 7, 8, 9, 10\}$. We denote this number by $\binom{10}{3}$. Let us suppose that we have a list of all these $\binom{10}{3}$ samples. Each sample contains three distinct numbers and each sample could be permuted in $3!$ different ways. Were we to do this, the result might look like Table 2.3, in which only some of the possible samples are listed; then each sample is permuted in all possible ways, so each sample gives $3!$ permutations.

There are two ways in which to view the contents of the table, which, if shown in its entirety, would contain all the permutations of 10 objects taken 3 at a time.

Table 2.3

Sample	Permutations					
{1,4,7}	1,4,7	1,7,4	4,1,7	4,7,1	7,1,4	7,4,1
{2,4,9}	2,4,9	2,9,4	4,2,9	4,9,2	9,2,4	9,4,2
{6,7,10}	6,7,10	6,10,7	7,6,10	7,10,6	10,6,7	10,7,6
⋮	⋮	⋮	⋮	⋮	⋮	⋮

First, using our formula for the number of permutations of 10 objects taken 3 at a time, the table must contain $\frac{10!}{(10-3)!}$ permutations. However, since each of the $\binom{10}{3}$ combinations can be permuted in 3! ways, the total number of permutations must also be $3! \cdot \binom{10}{3}$. It then follows that

$$3! \cdot \binom{10}{3} = \frac{10!}{(10-3)!}$$

From this, we see that

$$\binom{10}{3} = \frac{10!}{7! \cdot 3!} = \frac{10 \cdot 9 \cdot 8 \cdot 7!}{7! \cdot 3 \cdot 2 \cdot 1} = 120$$

This process is easily generalized. If we have $\binom{n}{r}$ distinct samples, each of these can be permuted in $r!$ ways, yielding all the permutations of n objects taken r at a time, so

$$r! \cdot \binom{n}{r} = \frac{n!}{(n-r)!}$$

or

$$\binom{n}{r} = \frac{n!}{r! \cdot (n-r)!}$$

$\binom{52}{5}$ then represents the total number of possible poker hands. This is 2, 598, 960. This number is small enough so that one could, given enough time, enumerate each of these.

This calculation by hand would appear this way:

$$\binom{52}{5} = \frac{52!}{5!(52-5)!} = \frac{52!}{5!47!} = \frac{52 \cdot 51 \cdot 50 \cdot 49 \cdot 48 \cdot 47!}{5 \cdot 4 \cdot 3 \cdot 2 \cdot 1 \cdot 47!}$$
$$= \frac{52 \cdot 51 \cdot 50 \cdot 49 \cdot 48}{5 \cdot 4 \cdot 3 \cdot 2 \cdot 1} = 2, 598, 960$$

Notice that the factors of 47! cancel from both the numerator and the denominator of the fraction above. This cancellation always occurs and a calculation rule is that $\binom{n}{r}$ has r factors in the numerator and in the denominator, so

$$\binom{n}{r} = \frac{n(n-1)(n-2) \cdots [n-(r-1)]}{r!}$$

This makes the calculation by hand fairly simple. We will solve some interesting problems after stating and proving some facts about these binomial coefficients.

It is also true that

$$\binom{n}{r} = \binom{n}{n-r}$$

An easy way to see this is to notice that if r objects are selected from n objects, then $n-r$ objects are left unselected. Every change in an item selected produces a change in the set of items that are not selected, and so the equality follows.

It is also true that $\binom{n}{r} = \binom{n-1}{r-1} + \binom{n-1}{r}$. To prove this, suppose you are a member of a group of n people and that a committee of size r is to be selected. Either you are on the committee or you are not on the committee. If you are selected for the committee, there are $\binom{n-1}{r-1}$ further choices to be made. There are $\binom{n-1}{r}$ committees that do not include you. So $\binom{n}{r} = \binom{n-1}{r-1} + \binom{n-1}{r}$.

Many other facts are known about the numbers $\binom{n}{r}$, which are also called *binomial coefficients* because they occur in the binomial theorem. The binomial theorem states that

$$(a+b)^n = \sum_{i=1}^{n} \binom{n}{i} a^{n-i} b^i$$

This is fairly easy to see. Consider $(a+b)^n = (a+b) \cdot (a+b) \cdot \cdots \cdot (a+b)$, where there are n factors on the right-hand side. To find the product $(a+b)^n$, we must choose either a or b from each of the factors on the right-hand side. There are $\binom{n}{i}$ ways to select i b's (and hence $n-i$ a's). The product $(a+b)^n$ consists of the sum of all such terms.

Many other facts are known about the binomial coefficients. We cannot explore all these here, but we will show an application, among others, to acceptance sampling.

EXAMPLE 2.3 *Arranging Some Like Objects*

Let us return to the problem first encountered when we counted the permutations of objects, some of which are alike. Specifically, we wanted to count the number of distinct permutations of 3 A's, 4 B's, and 5 C's, where the individual letters are not distinguishable from one another. We found the answer was $G = \frac{12!}{5!4!3!} = 27,720$.

Here's another way to arrive at the answer.

From the 12 positions in the permutation, choose 3 for the A's. This can be done in $\binom{12}{3}$ ways. Then from the remaining nine positions, choose four for the B's. This can be done in $\binom{9}{4}$

ways. Finally, there are five positions left for the 5 C's. So the total number of permutations must be $\binom{12}{3} \cdot \binom{9}{4} \cdot \binom{5}{5}$, which can be simplified to $\frac{12! \cdot 9! \cdot 5!}{3! \cdot 9!4! \cdot 5!5! \cdot 0!} = \frac{12!}{5!4!3!}$, as before. ∎

Note that we have used *combinations* to count permutations!

General Addition Theorem and Applications

In Chapter 1, we discussed some properties of probabilities including the addition theorem for two events: $P(A \cup B) = P(A) + P(B) - P(A \cap B)$. What if we have three or more events? This addition theorem can be generalized and we call this, following Chapter 1,

Fact 3. (*General addition theorem*).

$$P(A_1 \cup A_2 \cup \cdots \cup A_n) = \sum_{i=n}^{n} P(A_i) - \sum_{i \neq j=1}^{n} P(A_i \cap A_j) +$$

$$\sum_{i \neq j \neq k=1}^{n} P(A_i \cap A_j \cap A_k) - \cdots \pm \sum_{i \neq j \neq \cdots \neq n=1}^{n} P(A_i \cap A_j \cap \cdots \cap A_n)$$

We could not prove this in Chapter 1 since our proof involves combinations.

To prove the general addition theorem, we use a different technique from the one we used to prove the theorem for two events. Suppose a sample point is contained in exactly k of the events A_i. For convenience, number the events so that the sample point is in the first k events. Now we show that the probability of the sample point is contained exactly once in the right-hand side of the theorem.

The point is contained on the right-hand side

$$\binom{k}{1} - \binom{k}{2} + \binom{k}{3} - \cdots \pm \binom{k}{k}$$

times. But consider the binomial expansion of

$$0 = [1 + (-1)]^k = 1^k - \binom{k}{1} + \binom{k}{2} - \binom{k}{3} + \cdots \mp \binom{k}{k}$$

which shows that

$$\binom{k}{1} - \binom{k}{2} + \binom{k}{3} - \cdots \pm \binom{k}{k} = 1$$

So the sample point is counted exactly once, proving the theorem.

The principle we used here is that of *inclusion and exclusion* and is of great importance in discrete mathematics. It could also have been used in the case $k = 2$.

EXAMPLE 2.4   *Checking Hats*

Now we return to Example 1.2, where n diners have checked their hats and we seek the probability that at least one diner is given his own hat at the end of the evening. Let the events A_i denote the event "diner i gets his own hat," so we seek $P(A_1 \cup A_2 \cup \cdots \cup A_n)$ using the general addition theorem.

Suppose diner i gets his own hat. There are $(n-1)!$ ways for the remaining hats to be distributed, given the correct hat to diner i, so $P(A_i) = (n-1)!/n!$. There are $\binom{n}{1}$ ways for a single diner to be selected.

In a similar way, if diners i and j get their own hats, the remaining hats can be distributed in $(n-2)!$ ways, so $P(A_i \cap A_j) = (n-2)!/n!$. There are $\binom{n}{2}$ ways for two diners to be chosen. Clearly, this argument can be continued. We then find that

$$P(A_1 \cup A_2 \cup \cdots \cup A_n) = \binom{n}{1}\frac{(n-1)!}{n!} - \binom{n}{2}\frac{(n-2)!}{n!}$$
$$+ \binom{n}{3}\frac{(n-3)!}{n!} - \cdots \pm \binom{n}{n}\frac{(n-n)!}{n!}$$

which simplifies easily to

$$P(A_1 \cup A_2 \cup \cdots \cup A_n) = \frac{1}{1!} - \frac{1}{2!} + \frac{1}{3!} - \cdots \pm \frac{1}{n!}$$

Table 2.4 shows some numerical results from this formula.

Table 2.4

n	p
1	1.00000
2	0.50000
3	0.66667
4	0.62500
5	0.63333
6	0.63194
7	0.63214
8	0.63212
9	0.63212

It is perhaps surprising that, while the probabilities fluctuate a bit, they appear to approach a limit. To six decimal places, the probability that at least one diner gets his own hat is 0.632121 for $n \geq 9$. It can also be shown that this limit is $1 - 1/e$ for $n \geq 9$. ∎

EXAMPLE 2.5   *Aces and Kings*

Now we can solve the problem involving a real drug and a placebo given at the beginning of Chapter 1. To make an equivalent problem, suppose we seek the probability that when the cards are turned up one at a time in a shuffled deck of 52 cards all the aces will turn up before any of the kings. This is the same as finding the probability all the users of the real drug will occur before any of the users of the placebo.

The first insight into the card problem is that the remaining 44 cards have absolutely nothing to do with the problem. We need to only concentrate on the eight aces and kings.

Assume that the aces are indistinguishable from one another and that the kings are indistinguishable from one another. There are then $\binom{8}{4} = 70$ possible orders for these cards; only one of them has all the aces preceding all the kings, so the probability is $1/70$. ■

EXAMPLE 2.6   *Poker*

We have seen that there are $\binom{52}{5} = 2,598,960$ different hands that can be dealt in playing poker. We will calculate the probabilities of several different hands. We will see that the special hands have very low probabilities of occurring.

Caution is advised in calculating the probabilities: choose the values of the cards first and then the actual cards. Order is not important. Here are some of the possible hands and their probabilities.

(a) *Royal flush.* This is a sequence of 10 through ace in a single suit. Since there are four of these, the probability of a royal flush is $4/\binom{52}{5} = 0.000001539$.

(b) *Four of a kind.* This hand contains all four cards of a single value plus another card that must be of another value. Since there are 13 values to choose from for the four cards of a single value (and only one way to select them) and then 12 possible values for the fifth card, and then 4 choices for a card of that value, the probability of this hand is $13 \cdot 12 \cdot 4/\binom{52}{5} = 0.0002401$.

(c) *Straight.* This is a sequence of five cards regardless of suit. There are nine possible sequences, 2 through 6, 3 through 7, ..., 10 through Ace, and since the suits are not important, there are nine possible sequences and four choices for each of the five cards in the sequence, the probability of a straight is $9 \cdot 4^5/\binom{52}{5} = 0.003\,546$.

(d) *Two pairs.* There are $\binom{13}{2}$ choices for the values of the pairs and then $\binom{4}{2}$ choices for the two cards in the first pair and $\binom{4}{2}$ choices for the two cards in the second pair. Finally, there are 11 choices for the value of the fifth card and 4 choices for that card. So the probability of two pairs is $\binom{13}{2} \cdot \binom{4}{2} \cdot \binom{4}{2} \cdot 11 \cdot 4/\binom{52}{5} = 0.04754$.

(e) *Other special hands are* three of a kind (three cards of one value and two other cards of different values), full house (one pair and one triple), and one pair. The probabilities of these hands are $0.02113, 0.0014406$, and 0.422569, respectively. The most common hand is the one with five different values. This has probability $\binom{13}{5} \cdot 4^5/\binom{52}{5} = 0.507\,083$. The probability of a hand with at least one pair is then $1 - 0.507\,083 = 0.492\,917$. ■

Now we show another example, the one concerning auto racing.

EXAMPLE 2.7   *Race Cars*

One hundred race cars, numbered from 1 to 100, are running around a race course. We observe a sample of five noting the numbers on the cars and then calculate the median (the number in the middle when the sample is arranged in order). If the median is m, then we must choose two that are less than m and then two that are greater than m. This can be done in $\binom{m-1}{2} \cdot \binom{100-m}{2}$ ways. So the probability that the median is m is

$$\frac{\binom{m-1}{2} \cdot \binom{100-m}{2}}{\binom{100}{5}}$$

A graph of this function of m is shown in Figure 2.2.
The most likely value of m is 50 or 51, each having probability 0.0191346.

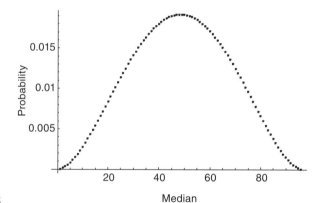

Figure 2.2

The race car problem is hardly a practical one. A more practical problem is this; we have taken a random sample of size 5 and we find that the median of the sample is 8. How many cars are racing around the track; that is, what is n?

This problem actually arose during World War II. The Germans numbered all kinds of war materiel and their parts. When we captured some tanks, say, we could then estimate the total number of tanks they had from the serial numbers on the captured tanks.

Here we will consider *maximum likelihood* estimation: we will estimate n as the value that makes the sample median we observed most likely.

If there are n tanks, then the probability the median of a sample of 5 tanks is m is

$$\frac{\binom{m-1}{2} \cdot \binom{n-m}{2}}{\binom{n}{5}}$$

Table 2.5

n	Probability
10	0.0833333
11	0.136364
12	0.159091
13	0.16317
14	0.157343
15	0.146853
16	0.134615
17	0.122172
18	0.110294
19	0.0993292
20	0.0893963
21	0.0804954
22	0.0725678
23	0.0655294
24	0.0592885
25	0.0537549

Now let us compute a table of values of n and these probabilities, letting $m = 8$ for various values of n. This is shown in Table 2.5.

A graph is helpful (Figure 2.3).

We see that the maximum probability occurs when $n = 13$, so we have found the maximum likelihood estimator for n. The mathematical solution of this problem would be a stretch for most students in this course. The computer is of great value here in carrying out a fairly simple idea.

It should be added here that this is not the optimal solution for the German tank problem. It should be clear that the maximum value in the sample carries more information about n than does the median. We will return to this problem in Chapter 12.   ∎

Now we return to the executive who is selecting the best assistant.

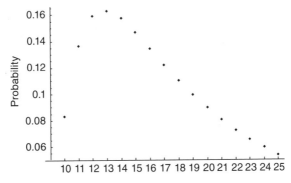

Figure 2.3

EXAMPLE 2.8   *Choosing an Assistant*

An executive in a company has an opening for an executive assistant. Twenty candidates have applied for the position. The executive is constrained by company rules that say that candidates must be told whether they are selected or not at the time of an interview. How should the executive proceed so as to maximize the chance that the best candidate is selected?

We have already seen that a random choice of candidate selects the best one with probability $1/20 = 0.05$, so it is not a very sensible strategy.

It is probably clear, assuming the candidates appear in random order, that we should deny the job to a certain number of candidates while noting which one was best in this first group; then we should choose the next candidate who is better than the best candidate in the first group (or the last candidate if a better candidate does not appear).

This strategy has surprising consequences. To illustrate what follows from it, let us consider a small example of four candidates whom we might as well number 1, 2, 3, 4 with 4 being the best candidate and 1 the worst candidate. The candidates can appear in $4! = 24$ different orders that are shown in Table 2.6.

Table 2.6

Order	Pass 1	Pass2	Order	Pass 1	Pass 2
1234	2	3	3124	4	4
1243	2	4	3142	4	4
1324	3	4	3241	4	4
1342	3	4	3214	4	4
1423	4	3	3412	4	2
1432	4	2	3421	4	1
2134	3	3	4123	3	3
2143	4	4	4132	2	2
2314	3	4	4231	1	1
2341	3	4	4213	3	3
2413	4	3	4312	2	2
2431	4	1	4321	1	1

The column headed "*Pass* 1" indicates the final choice when the first candidate is passed by and the next candidate ranking higher than the first candidate is selected.

Similarly, the column headed "*Pass* 2" indicates the final choice when the first two candidates are passed by and the next candidate ranking higher than the first two candidates is selected.

It is only sensible to pass by one or two candidates since we will choose the fourth candidate if we let three candidates pass by and the probability of choosing the best candidate is then $1/4$.

So we let one or two candidates pass, noting the best of these. Then we choose the next candidate better than the best in the group we passed. Suppose we interview one candidate, reject him or her, noting his or her ranking. So if the candidates appeared in the order 3214, then we would pass by the candidate ranked 3; the next best candidate is 4. These rankings appear in Table 2.6 under the column labeled "*Pass* 1". If we examine the rankings and their frequencies in that column, we get Table 2.7.

Table 2.7 Passing the First Candidate

Ranking	Frequency
1	2
2	4
3	7
4	11

Interestingly, the most frequent choice is 4! and the average of the rankings is 3.125, so we do fairly well.

If we interview the first two candidates, noting the best, and then choose the candidate with the better ranking (or the last candidate), we find the rankings in Table 2.6 labeled "*Pass 2.*" A summary of our choice is shown in Table 2.8.

Table 2.8 Passing the First Two Candidates

Ranking	Frequency
1	4
2	4
3	6
4	10

We do somewhat less well, but still better than a random choice. The average ranking here is 2.917.

A little forethought would reduce the number of permutations we have to list. Consider the plan to pass the first candidate by. If 4 appears in the first position, we will not choose the best candidate; if 4 appears in the second position, we will choose the best candidate; if 3 appears in the first position, we will choose the best candidate; so we did not need to list 17 of the 24 permutations. Similar comments will apply to the plan to pass the first two candidates by.

It is possible to list the permutations of five or six candidates and calculate the average choice; beyond that this procedure is not very sensible. The results for five candidates are shown in Table 2.9, where the first candidate is passed by.

Table 2.9 Five Candidates Passing the First One

Ranking	Frequency
1	6
2	12
3	20
4	32
5	50

The plan still does very well. The average rank selected is 3.90 and we see with the plans presented here that we get the highest ranked candidate at least 42% of the time.

Generalizing the plan, however, is not so easy. It can be shown that the optimal plan passes the first $[n/e]$ candidates (where the brackets indicate the greatest integer function) and the probability of selecting the best candidate out of the n candidates approaches $1/e$ as n increases. So we allow the first candidate to pass by until $n = 6$ and then let the first two candidates pass by until $n = 9$, and so on. ■

This problem makes an interesting classroom exercise that is easily simulated with a computer that can produce random permutations of n integers.

EXAMPLE 2.9 *Acceptance Sampling*

Now let us discuss an acceptance sampling plan.

Suppose a lot of 100 items manufactured in an industrial plant actually contains items that do not meet either the manufacturer's or the buyer's specifications. Let us denote these items by calling them D items while the remainder of the manufacturer's output, those items that do meet the manufacturer's and the buyer's specifications, we will call G items.

Now the manufacturer wishes to inspect a random sample of the items produced by the production line. It may be that the inspection process destroys the product or that the inspection process is very costly, so the manufacturer uses sampling and so inspects only a portion of the manufactured items.

As an example, suppose the lot of 100 items actually contains 10 D items and 90 G items and that we select a random sample of 5 items from the entire lot produced by the manufacturer. There are $\binom{100}{5} = 75,287,520$ possible samples. Suppose we want the probability that the sample contains exactly three of the D items. Since we assume that each of the samples is equally likely, this probability is

$$P(D = 3) = \frac{\binom{10}{3} \cdot \binom{90}{2}}{\binom{100}{5}}$$

$$= 0.00638353$$

making it fairly unlikely that this sample will find three of the items that do not meet specifications.

It may be of interest to find the probabilities for all the possible values of D. This is often called the *probability distribution* of the *random variable D*. That is, we want to find the values of the function

$$f(d) = \frac{\binom{10}{d} \cdot \binom{90}{5-d}}{\binom{100}{5}}$$

for $d = 0, 1, 2, 3, 4, 5$.

A graph of this function is shown in Figure 2.4.

What should the manufacturer do if items not meeting specifications are discovered in the sample? Normally, one of two courses is followed: either the D items found in the sample are

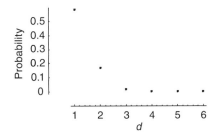

Figure 2.4

replaced by G items or the entire lot is inspected and any D items found in the entire lot are replaced by G items. The last course is usually followed if the sample does not exhibit too many D items, and, of course, can only be followed if the sampling is not destructive.

If the sample does not contain too many D items, the lot is *accepted* and sent to the buyer, perhaps after some D items in the sample are replaced by G items. Otherwise, the lot is *rejected*. Hence,the process is called *acceptance sampling*.

We will explore the second possibility noted above here, namely, that if any D items at all are found in the sample, then the entire lot is inspected and any D items in it are replaced with G items. So, the entire delivered lot consists of G items when the sample detects any D items at all. This clearly will improve the quality of the lot of items sold, but it is not clear how much of an improvement will result. The process has some surprising consequences and we will now explore this procedure.

To be specific, let us suppose that the lot is accepted only if the sample contains no D items whatsoever. Let us also assume that we do not know how many D items are in the lot, so we will suppose that there are d of these in the lot.

The lot is then accepted with probability

$$P(D=0) = \frac{\binom{100-d}{5}}{\binom{100}{5}}$$

This is a decreasing function of d; the larger d, is, the more likely the sample will contain some D items and hence the lot will not be accepted. A graph of this function is shown in Figure 2.5.

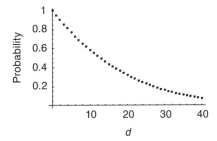

Figure 2.5

Finally, we consider the *average percentage of D items* delivered to the customer with this acceptance sampling plan. This is often called the *average outgoing quality* (AOQ) in the quality control literature.

The average of a quantity is found by multiplying the values of that quantity by the probability of that quantity and adding the results. So if a random variable is D, say, whose specific values are d, then the average value of D is

$$\sum_{\text{all values of } d} d \cdot P(D = d)$$

Here we wish to find the average value of the percentage of D items delivered to the buyer, or the average of the quantity $d/100$. This is the average outgoing quality.

$$\text{AOQ} = \sum_{\text{all values of } d} \frac{d}{100} \cdot P(D = d)$$

But we have a very special circumstance here. The delivered lot has percentage D items of $d/100$ only if the sample contains no D items whatsoever; otherwise, the lot has 0% D items due to the replacement plan. So the average outgoing quality is

$$\text{AOQ} = \frac{d}{100} \cdot P(D = 0) + \frac{0}{100} \cdot P(D \neq 0)$$

so

$$\text{AOQ} = \frac{d}{100} \cdot P(D = 0)$$

or

$$\text{AOQ} = \frac{d}{100} \cdot \frac{\binom{100 - d}{5}}{\binom{100}{5}}$$

A graph of this function is shown in Figure 2.6.

We notice that the graph attains a maximum value; this may not have been anticipated! This means that regardless of the quality of the lot, there is a maximum for the average percentage of D items that can be delivered to the customer! This maximum can be found using a computer and the above graph. Table 2.10 shows the values of the AOQ near the maximum value.

We see that the maximum AOQ occurs when $d = 16$, so the maximum average percentage of D items that can be delivered to the customer is 0.066!

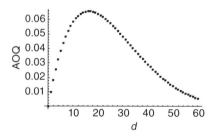

Figure 2.6

Table 2.10

d	AOQ
14	0.06476
15	0.06535
16	0.06561
17	0.06556
18	0.06523

Sampling here has had a dramatic impact on the average percentage of D items delivered to the customer.   ■

This is just one example of how probability and statistics can assist in delivering high-quality product to consumers. There are many other techniques used that are called in general statistical process control methods, or SPC; these have found wide use in industry today. Statistical process control is the subject of Chapter 11.

CONCLUSIONS

We have explored permutations and combinations in this chapter and have applied them to several problems, most notably a plan for choosing the best candidate from a group of applicants for a position and acceptance sampling where we found that sampling does improve the quality of the product sold and actually puts a limit on the percentage of unacceptable product sold.

We will continue our discussion or production methods and the role probability and statistics can play in producing more acceptable product in the chapter on quality control and statistical process control.

We continue in the following two chapters with a discussion of conditional probability and geometric probability. Each of these topics fits well into a course in geometry.

EXPLORATIONS

1. Use the principle of inclusion and exclusion to prove the general addition theorem for two events.
2. Find the probability a poker hand has
   (a) exactly two aces;
   (b) exactly one pair.
3. What is the probability a bridge hand (13 cards from a deck of 52 cards) does not contain a heart?
4. Simulate Example 2.8: Choose random permutations of the integers from 1 to 10 with 10 being the most qualified assistant. Compare the probabilities

of making the best choice by letting 1, 2, or 3 applicants go by and then choosing the best applicant better than the best of those passed by.

5. *Simulate Example* 2.4: Choose random permutations of the integers from 1 to 5 and count the number of instances of an integer being in its own place. Then count the number of derangements, that is, permutations where no integer occupies its own place.

6. In Example 2.7, how would you use the maximum of the sample in order to estimate the maximum of the population?

7. A famous correspondence between the Chevalier de Mere and Blaise Pascal, each of whom made important contributions to the theory of probability, contained the question. "Which is more likely—at least one 6 in 4 rolls of a fair die or at least one sum of 12 in 24 rolls of a pair of dice?"

Show that the two events are almost equally likely. Which is more likely?

Chapter 3

Conditional Probability

CHAPTER OBJECTIVES:

- to consider some problems involving *conditional probability*
- to show diagrams of conditional probability problems
- to show how these probability problems can be solved using only the area of a rectangle
- to show connections with geometry
- to show how a test for cancer and other medical tests can be misinterpreted
- to analyze the famous Let's Make a Deal problem.

INTRODUCTION

A physician tells a patient that a test for cancer has given a positive response (indicating the possible presence of the disease). In this particular test, the physician knows that the test is 95% accurate both for patients who have cancer and for patients who do not have cancer. The test appears at first glance to be quite accurate. How is it then, based on this test alone, that this patient almost certainly does not have cancer? We will explain this.

I have a class of 65 students. I regard one of my students as an expert since she never makes an error. I regret to report that the remaining students are terribly ignorant of my subject, and so guess at each answer. I gave a test with six true–false questions; a paper I have selected at random has each of these questions answered correctly. What is the probability that I have selected the expert's paper? The answer may be surprising.

Each of these problems can be solved using what is usually known in introductory courses as *Bayes' theorem*. We will not need this theorem at all to solve these problems. No formulas are necessary, except that for the area of a rectangle, so these problems can actually be explained to many students of mathematics.

These problems often cause confusion since it is tempting to interchange two probabilities. In the cancer example, the probability that the test is positive if the patient has cancer (0.95 in our example) is quite different from the probability that the patient has cancer if the test is positive (only 0.087!). We will explain how this apparent difference arises and how to calculate the conditional probability, which is probably the one of interest.

Finally, and regrettably, in the second example, despite the large size of my class, the probability I am looking at the expert's paper is only 1/2. Now let us see how to tackle these problems in a simple way. We begin with two simple examples and then solve the problems above. We will also discuss the famous Let's Make a Deal problem.

EXAMPLE 3.1   *For the Birds*

A pet store owner purchases bird seed from two suppliers, buying 3/4 of his bird seed from one supplier and the remainder from another. The germination rate of the seed from the first supplier is 75% whereas the germination rate from the second supplier is 60%. If seed is randomly selected and germinates, what is the probability that it came from the first supplier?

We will show how to make this problem visual and we will see that it is fairly simple to analyze. We begin with a square of side 1, as shown in Figure 3.1. The total area of the square is 1, so portions of the total area represent probabilities.

Along the horizontal axis we have shown the point at 3/4, indicating the proportion of seed purchased from the first supplier. The remainder of the axis is of length 1/4, indicating the proportion of the seed purchased from the second supplier.

Along the vertical axis, we have shown a horizontal line at 75%, the proportion of the first supplier's seed that germinates. The area of the rectangle formed is $3/4 \cdot 0.75$, which is the probability that the seed came from the first supplier and germinates. This is shaded in Figure 3.1. We have also shown a horizontal line at the 60% mark along the vertical axis, indicating that this is the percentage of the second supplier's seed that germinates. The area of the unshaded rectangle with height 0.60 and whose base is along the horizontal axis is $1/4 \cdot 0.60$,

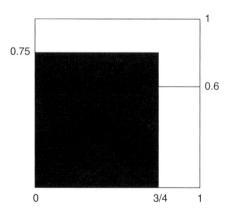

Figure 3.1

representing the probability that the seed came from the second supplier and germinates. Now the total area of the two rectangles is $3/4 \cdot 0.75 + 1/4 \cdot 0.60 = 0.7125$; this is the probability that the seed germinates regardless of the supplier.

We now want to find the probability that germinating seed came from the first supplier. This is the portion of the two rectangles that is shaded or

$$\frac{3/4 \cdot 0.75}{3/4 \cdot 0.75 + 1/4 \cdot 0.60} = 0.789474$$

differing from $3/4$, the percentage from the first supplier.   ∎

This problem and the others considered in this chapter are known as *conditional probability* problems. They are usually solved using *Bayes' theorem*, but, as we have shown above, all that is involved is areas of rectangles.

We emphasize here that the *condition* is the crucial distinction. It matters greatly whether we are given the condition that the seed came from the first supplier (in which case the germination rate is 75%) or that the seed germinated (in which case the probability it came from the first supplier is 78.474%). Confusing these two conditions can lead to erroneous conclusions, as some of our examples will show.

EXAMPLE 3.2   *Driver's Ed*

In a high school, 60% of the students take driver's education. Of these, 4% have an accident in a year. Of the remaining students, 8% have an accident in a year. A student has an accident; what is the probability he or she took driver's ed?

Note that the probability a driver's ed student has an accident (4%) may well differ from the probability he took driver's ed if he had an accident. Let us see if this is so. The easiest way to look at this is again to draw a picture. In Figure 3.2 we show the unit square as we did in the first example.

Along the horizontal axis, we have shown 60% of the axis for students who take driver's ed while the remainder of the horizontal axis, 40%, represents students who do not take driver's education. Now of the 60% who take driver's ed, 4% have an accident, so we show a line on the vertical axis at 4%.

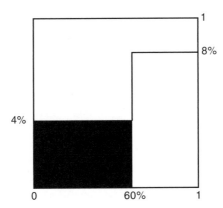

Figure 3.2

The area of this rectangle, $0.6 \cdot 0.04 = 0.024$, represents students who both take driver's ed and have an accident. (The scale has been exaggerated here so that the rectangles we are interested in can be seen more easily.) This rectangle has been shaded.

Above the 40% who do not take driver's ed, we show a line at 8%.

This rectangle on the right has area $0.4 \cdot 0.08 = 0.032$, representing students who do not take driver's ed and who have an accident. This is unshaded in the figure.

The area of the two rectangles then represents the proportion of students who have an accident. This is $0.024 + 0.032 = 0.056$.

Now we want the probability that a student who has an accident has taken driver's ed. This is the area of the two rectangles that arises from the left-hand rectangle, or $0.024/0.056 = 3/8 = 37.5\%$. It follows that $5/8 = 62.5\%$ of the students who have accidents did not take driver's ed.

It is clear then that the probability a student has an accident if he took driver's ed (4%) differs markedly from the probability a student who took driver's ed had an accident (37.5%).   ∎

It is not uncommon for these probabilities to be mixed up. Note that the first relates to students who have had accidents whereas the second refers to the group of students who took driver's ed.

SOME NOTATION

Now we pause in our examples to consider some notation.

From this point on, we will use the notation, as we have done previously, $P(A|B)$ to denote "the probability that an event A occurs given that an event B has occurred."

In the bird seed example, we let $S1$ and $S2$ denote the suppliers and let G denote the event that the seed germinates.

We then have $P(G|S1) = 0.75$ and $P(G|S2) = 0.60$, and we found that $P(S1|G) = 0.78474$. Note that the condition has a great effect on the probability.

EXAMPLE 3.3   *The Expert in My Class*

Recall that our problem is this: I have a class of 65 students. I regard one of my students as an expert since she never makes an error. I regret to report that the remaining students are terribly ignorant of my subject, and so guess at each answer. I gave a test with six true–false questions; a paper I have selected at random has each of these questions answered correctly.

What is the probability that I have selected the expert's paper? One might think this is certain, but it is not.

This problem can be solved in much the same way that we solved the previous two problems.

Begin again with a square of side 1. On the horizontal axis, we indicate the probability we have chosen the expert (1/65) and the probability we have chosen an ordinary student (64/65). The chance of choosing the expert is small, so we have exaggerated the scale in Figure 3.3 for reasons of clarity. The relative areas then should not be judged visually.

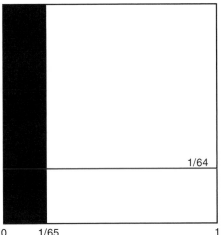

Figure 3.3   0   1/65   1

Now the expert answers the six true–false questions on my examination with probability 1. This has been indicated on the vertical axis. The area of the corresponding rectangle is then

$$\frac{1}{65} \cdot 1 = \frac{1}{65}$$

This represents the probability that I chose the expert and that all the questions are answered correctly.

Now the remainder of my students are almost totally ignorant of my subject and so they guess the answers to each of my questions. The probability that one of these students answers all the six questions correctly is then $(1/2)^6 = 1/64$. This has been indicated on the right-hand vertical scale in Figure 3.3. The rectangle corresponding to the probability that a nonexpert answers all six of the questions correctly is then

$$P(A|\overline{E}) = \frac{64}{65} \cdot \frac{1}{64} = \frac{1}{65}$$

Let us introduce some notation here. Let E denote the expert and $\overline{E}$ denote a nonexpert and let A denote the event that the student responds correctly to each question. We found that $P(A|E) = 1$ while we want $P(E|A)$, so again the condition is crucial.

So the total area, shown as two rectangles in Figure 3.3, corresponding to the probability that all six of the questions are answered correctly, is

$$P(A) = P(A \text{ and } E) + P(A \text{ and } \overline{E})$$
$$P(A) = \frac{1}{65} \cdot 1 + \frac{64}{65} \cdot \frac{1}{64}$$
$$= \frac{2}{65}$$

The portion of this area corresponding to the expert is

$$P(E|A) = \frac{\frac{1}{65} \cdot 1}{\frac{1}{65} \cdot 1 + \frac{64}{65} \cdot \frac{1}{64}}$$

$$= \frac{1}{2}$$

The shaded areas in Figure 3.3 are in reality equal, but they do not appear to be equal.   ∎

Recall that the statement $P(A) = P(A \text{ and } E) + P(A \text{ and } \overline{E})$ is often known as the *Law of Total Probability*.

We add the probabilities here of mutually exclusive events, namely, A and E and A and $\overline{E}$ (represented by the nonoverlapping rectangles). The law can be extended to more than two mutually exclusive events.

EXAMPLE 3.4   *The Cancer Test*

We now return to the medical example at the beginning of this chapter. We interpret the statement that the cancer test is accurate for 95% of patients with cancer and 95% accurate for patients without cancer as that $P(T^+|C) = 0.95$ and $P(T^+|\overline{C}) = 0.05$, where C indicates the presence of cancer and T^+ means that the test indicates a positive result or the presence of cancer. $P(T^+|\overline{C})$ is known as the *false positive rate* for the test since it produces a positive result for noncancer patients. Supposing that a small percentage of the population has cancer, we assume in this case that $P(C) = 0.005$. This assumption will prove crucial in our conclusions.

A picture will clarify the situation, although, again, the small probabilities involved force us to exaggerate the picture somewhat. Figure 3.4 shows along the horizontal axis the probability that a person has cancer. Along the vertical axis are the probabilities that a person shows a positive test for each of the two groups of patients.

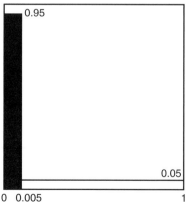

Figure 3.4   0   0.005   1

It is clear that the probability that a person shows a positive test is

$$P(T^+) = 0.95 \cdot 0.005 + 0.05 \cdot 0.995$$

$$= 0.0545$$

The portion of this area corresponding to the people who actually have cancer is then

$$P(C|T^+) = \frac{0.95 \cdot 0.005}{0.95 \cdot 0.005 + 0.05 \cdot 0.995}$$

$$= \frac{0.00475}{0.0545} = 0.087$$

This is surprisingly low. We emphasize, however, that the test should not be relied upon alone; one should have other indications of the disease as well.

We also note here that the probability that a person testing positive actually has cancer highly depends upon the true proportion of people in the population who are actually cancer patients. Let us suppose that this true proportion is r, so that r represents the incidence rate of the disease. Replacing the proportion 0.005 by r and the proportion 0.995 by $1 - r$ in the above calculation, we find that the proportion of people who test positive and actually have the disease is

$$P(C|T^+) = \frac{r \cdot 0.95}{r \cdot 0.95 + (1 - r) \cdot 0.05}$$

This can be simplified to

$$P(C|T^+) = \frac{0.95 \cdot r}{0.05 + 0.90 \cdot r} = \frac{19r}{1 + 18r}$$

A graph of this function is shown in Figure 3.5.

We see that the test is quite reliable when the incidence rate for the disease is large. Most diseases, however, have small incidence rates, so the false positive rate for these tests is a very important number.

Now suppose that the test has probability p indicating the disease among patients who actually have the disease and that the test indicates the presence of the disease with probability $1 - p$ among patients who do not have the disease $p = 0.95$ in our example. It is also interesting

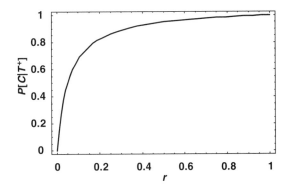

Figure 3.5

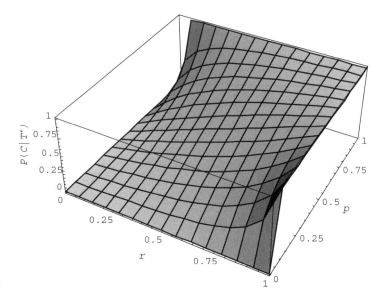

Figure 3.6

to examine the probability $P(C|T^+)$ as a function of both the incidence rate of the disease, r, and p. Now

$$P(C|T^+) = \frac{r \cdot p}{r \cdot p + (1-r) \cdot (1-p)}$$ ■

The surface showing this probability as a function of both r and p is shown in Figure 3.6.

Clearly, the accuracy of the test increases as r and p increase. Ours has a low probability since we are dealing with the lower left corner of the surface.

EXAMPLE 3.5   *Let's Make a Deal*

In this TV game show, a contestant is presented with three doors, one of which contains a valuable prize while the other two are empty. The contestant is allowed to choose one door. Regardless of the choice made, at least one, and perhaps two, of the remaining doors is empty. The show host, say Monty Hall, opens one door and shows that it is empty. He now offers the contestant the opportunity to change the choice of doors; should the contestant switch, or doesn't it matter?

It matters. The contestant who switches has probability 2/3 of winning the prize. If the contestant does not switch, the probability is 1/3 that the prize is won.

In thinking about the problem, note that when the empty door is revealed, the game does not suddenly become choosing one of the two doors that contains the prize. The problem here is that sometimes Monty Hall has one choice of door to show empty and sometimes he has two choices of doors that are empty. This must be accounted for in analyzing the problem.

An effective classroom strategy at this point is to try the experiment several times, perhaps using large cards that must be shuffled thoroughly before each trial; some students can use the "never switch" strategy whereas others can use the "always switch" strategy and the results compared. This experiment alone is enough to convince many people that the switching strategy is the superior one; we analyze the problem using geometry.

To be specific, let us call the door that the contestant chooses as door 1 and the door that the host opens as door 2. The symmetry of the problem tells us that this is a proper analysis of the general situation.

Now we need some notation. Let P_i, $i = 1, 2, 3$, denote the event "prize is behind door i" and let D be the event "door 2 is opened." We assume that $P_1 = P_2 = P_3 = 1/3$.

Then, $P(D|P_1) = 1/2$, since in that case the host then has a choice of two doors to open; $P(D|P_2) = 0$, since the host will not open the door showing the prize; and $P(D|P_3) = 1$, since in this case door 2 is the only one that can be opened to show no prize behind it.

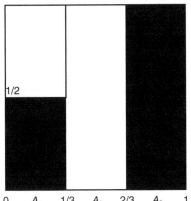

Figure 3.7 0 A_1 $1/3$ A_2 $2/3$ A_3 1

Our unit square is shown in Figure 3.7. It is clear that the shaded area in Figure 3.7 represents the probability that door 2 is opened. The probability that the contestant wins if he switches is then the proportion of this area corresponding to door 3. This is

$$P(P_3|D) = \frac{\frac{1}{3} \cdot 1}{\frac{1}{3} \cdot \frac{1}{2} + \frac{1}{3} \cdot 0 + \frac{1}{3} \cdot 1} = \frac{2}{3}$$ ∎

Another, perhaps more intuitive, way to view the problem is this: when the first choice is made, the contestant has probability $1/3$ of winning the prize. The probability that the prize is behind one of the other doors is $2/3$. Revealing one of the doors to be empty does not alter these probabilities; hence, the contestant should switch.

BAYES' THEOREM

Bayes' theorem is stated here although, as we have seen, problems involving it can be done geometrically.

Bayes' theorem: If $S = A_1 \cup A_2 \cup \cdots A_n$, where A_i and A_j have no sample points in common if $i \neq j$, then if B is an event,

$$P(A_i|B) = \frac{P(A_i \cap B)}{P(B)}$$

$$P(A_i|B) = \frac{P(A_i) \cdot P(B|A_i)}{P(A_1) \cdot P(B|A_1) + P(A_2) \cdot P(B|A_2) + \cdots + P(A_n) \cdot P(B|A_n)}$$

or

$$P(A_i|B) = \frac{P(A_i) \cdot P(B|A_i)}{\sum\limits_{j=1}^{n} P(A_j) \cdot P(B|A_j)}$$

In Example 3.5 (Let's Make a Deal), A_i is the event "Prize is behind door i" for $i = 1, 2, 3$. (We used P_i in Example 3.5.) B was the event "door 2 is opened."

CONCLUSIONS

The problems considered above should be interesting and practical for our students. I think our students should have mastery of these problems and others like them since data of this kind are frequently encountered in various fields.

Mathematically, the analyses given above are equivalent to those found by using a probability theorem known as Bayes' theorem. The geometric model given here shows that this result need not be known since it follows so simply from the area of a rectangle. The analysis given here should make these problems accessible to elementary students of mathematics.

EXPLORATIONS

1. Three methods, say A, B, and C, are sometimes used to teach an industrial worker a skill. The methods fail to instruct with rates of 20%, 10%, and 30%, respectively. Cost considerations restrict method A to be used twice as often as B that is used twice as often as C. If a worker fails to learn the skill, what is the probability that she was taught by method A?

2. Binary symbols (0 or 1) sent over a communication line are sometimes interchanged. The probability that a 0 is changed to 1 is 0.1 while the probability that a 1 is changed to 0 is 0.2. The probability that a 0 is sent is 0.4 and the probability that a 1 is sent is 0.6. If a 1 is received, what is the probability that a 0 was sent?

3. Sample surveys are often subject to error because the respondent might not truthfully answer a sensitive question such as "Do you use illegal drugs?" A procedure known as *randomized response* is sometimes used. Here is how that works. A respondent is asked to flip a coin and not reveal the result. If the coin comes up heads, the respondent answers the sensitive question, otherwise he

responds to an innocuous question such as "Is your Social Security number even?" So if the respondent answers "Yes," it is not known to which question he is responding. Show, however, that with a large number of respondents, the frequency of illegal drug use can be determined.

4. Combine cards from several decks and create another deck of cards with, say, 12 aces and 40 other cards. Have students select a card and not reveal whether it is an ace or not. If an ace is chosen, have the students answer the question about illegal drugs in the previous exploration and otherwise answer an innocuous question. Then approximate the use of illegal drugs.

5. A certain brand of lie detector is accurate with probability 0.92; that is, if a person is telling the truth, the detector indicates he is telling the truth with probability 0.92, while if he is lying, the detector indicates he is lying with probability 0.92. Assume that 98% of the subjects of the test are truthful. What is the probability that a person is lying if the detector indicates he is lying?

Chapter 4

Geometric Probability

CHAPTER OBJECTIVES:

- to use connections between geometry and probability
- to see how the solution of a quadratic equation solves a geometric problem
- to use linear inequalities in geometric problems
- to show some unusual problems for geometry.

EXAMPLE 4.1 *Meeting at the Library*

Joan and Jim agree to meet at the library after school between 3 and 4 p.m. Each agrees to wait no longer than 15 min for the other. What is the probability that they will meet?

This at first glance does not appear to be a geometric problem, but it is.

We show the situation in Figure 4.1. For convenience, we take the interval between 3 and 4 p.m. to be the interval between 0 and 1, so both Joan and Jim's waiting time becomes 1/4 of an hour. We suppose that each person arrives at some random time, so these arrival times are points somewhere in a square of side 1. Let X denote Joan's arrival time and Y denote Jim's arrival time.

If Joan arrives first, then Jim's arrival time Y must be greater than Joan's arrival time X. So $Y > X$ or $Y - X > 0$ and if they are to meet, then $Y - X < 1/4$ or $Y < X + 1/4$. This is the region below the line $Y = X + 1/4$ and has intercepts at $(0, 1/4)$ and $(3/4, 1)$ and is the top line in Figure 4.1.

If Jim arrives first, then $X > Y$ and $X - Y > 0$ and if they are to meet, then $X - Y < 1/4$ or $Y > X - 1/4$. This is the region above the line $Y = X - 1/4$ and has intercepts at $(1/4, 0)$ and $(1, 3/4)$ and is the lower line in Figure 4.1.

They both meet then when $Y < X + 1/4$ and when $Y > X - 1/4$. This is the region between the lines and is shaded in the figure.

Since the area of the square is 1, the shaded area must represent the probability that Joan and Jim meet. The easiest way to compute this is to subtract the areas of the two triangles

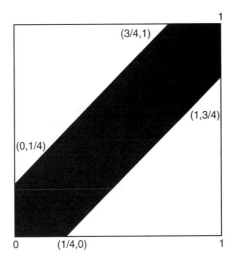

Figure 4.1

from 1, the area of the square. The triangles are equal in size. This gives us the probability that they meet as

$$P(\text{Joan and Jim meet}) = 1 - 2 \cdot \frac{1}{2} \cdot \left(\frac{3}{4}\right)^2 = 1 - \frac{9}{16} = \frac{7}{16} = 0.4375$$

So they meet less than half the time.   ∎

It is, perhaps, surprising that the solution to our problem is geometric. This is one of the many examples of the solution of probability problems using geometry. We now show some more problems of this sort.

EXAMPLE 4.2   *How Long Should They Wait?*

Now suppose that it is really important that they meet, so we want the probability that they meet to be, say, 3/4. How long should each wait for the other?

Let us say that each waits t minutes for the other. The situation is shown in Figure 4.2. We know then that the shaded area is 3/4 or that the nonshaded area is 1/4. This means that

$$1 - 2 \cdot \frac{1}{2} \cdot (1 - t)^2 = \frac{3}{4}$$

or that

$$1 - (1 - t)^2 = \frac{3}{4}$$

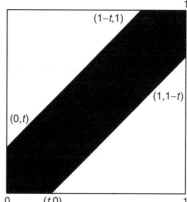

Figure 4.2

so

$$(1 - t)^2 = \frac{1}{4} \quad \text{or} \quad 1 - t = \frac{1}{2}$$

so

$$t = \frac{1}{2}$$

So each must wait up to 30 min for the other for them to meet with probability 3/4.   ■

EXAMPLE 4.3   *A General Graph*

In the previous example, we specified the probability that Joan and Jim meet. Now suppose we want to know how the waiting time, say t hours for each person, affects the probability that they meet.

The probability that they meet is

$$p = 1 - 2 \cdot \frac{1}{2} \cdot (1 - t)^2$$

or

$$t^2 - 2t + p = 0$$

    ■

Figure 4.3 is a graph of this quadratic function where t is restricted to be between 0 and 1.

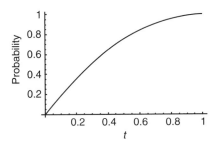

Figure 4.3

EXAMPLE 4.4   *Breaking a Wire*

I have a piece of wire of length L. I break it into three pieces. What is the probability that I can form a triangle from the three pieces of wire?

Suppose the pieces of wire after the wire is broken have lengths x, y, and $L - x - y$.

To form a triangle, the sum of the lengths of any two sides must exceed the length of the remaining side. So,

$$x + y > L - x - y \quad \text{or} \quad x + y > L/2$$

and

$$x + (L - x - y) > y \quad \text{so} \quad y < L/2$$

and

$$y + (L - x - y) > x \quad \text{so} \quad x < L/2$$

It is easy to see the simultaneous solution of these inequalities geometrically, as shown in Figure 4.4.

If $x < L/2$, then x must be to the left of the vertical line at $L/2$. If $y < L/2$, then y must be below the horizontal line at $L/2$. Finally, $x + y > L/2$; this is the region above the line $x + y = L/2$.

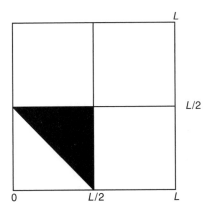

Figure 4.4

The resulting region is the triangle shaded in the figure. Its area is clearly 1/8 of the square; so the probability that I can form a triangle from the pieces of wire is 1/8.   ■

EXAMPLE 4.5   *Shooting Fouls*

A, B, and C are basketball players. A makes 40% of her foul shots, B makes 60% of her foul shots, and C makes 80% of her foul shots. A takes 50% of the team's foul shots, B takes 30% of the team's foul shots, and C takes 20% of the team's foul shots. What is the probability that a foul shot is made?

The situation can be seen in Figure 4.5, as we have done in all our previous examples.

The region labeled "A" is the region where player A is shooting and she makes the shot; the situation is similar for players B and C. So the probability that a shot is made is then the sum of the areas for the three players or

$$(0.5)(0.4) + (0.3)(0.6) + (0.2)(0.8) = 0.54$$

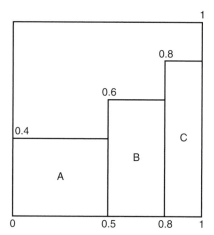

Figure 4.5

■

EXAMPLE 4.6   *Doing the Dishes*

We conclude this chapter with a problem that at first glance does not involve geometry. But it does!

My daughter, Kaylyn, and I share doing the dinner dishes. To make the situation interesting, I have in my pocket two red marbles and one green marble. We agree to select two marbles at random. If the colors match, I do the dishes; otherwise Kaylyn does the dishes. Is the game fair?

Of course the game is not fair. It is clear from the triangle in Figure 4.6 that Kaylyn does the dishes 2/3 of the time, corresponding to two sides of the triangle, while I do the dishes 1/3 of the time, corresponding to the base of the triangle.

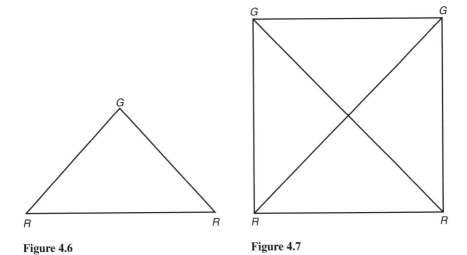

Figure 4.6   **Figure 4.7**

How can the game be made fair? It may be thought that I have too many red marbles in my pocket and that adding a green marble will rectify things.

However, examine Figure 4.7 where we have two green and two red marbles. There are six possible samples of two marbles; two of these contain marbles of the same color while four contain marbles of different colors. Adding a green marble does not change the probabilities at all!   ■

Why is this so? Part of the explanation lies in the fact that while the number of red marbles and green marbles is certainly important, it is the number of sides and diagonals of the figure involved that is crucial. It is the geometry of the situation that explains the fairness, or unfairness, of the game.

It is interesting to find, in the above example, that if we have three red marbles and one green marble, then the game is fair. The unfairness of having two red and one green marbles in my pocket did not arise from the presumption that I had too many red marbles in my pocket. I had too few!

Increasing the number of marbles in my pocket is an interesting challenge. Figure 4.8 shows three red and three green marbles, but it is not a fair situation.

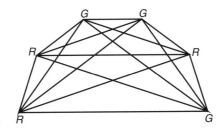

Figure 4.8

Table 4.1

R	G	R + G
3	1	4
6	3	9
10	6	16
15	10	25

The lines in Figure 4.8 show that there are 15 possible samples to be selected, 6 of which have both marbles of the same color while 9 of the samples contain marbles of different colors. With a total of six marbles, there is no way in which the game can be made fair.

Table 4.1 shows some of the first combinations of red and green marbles that produce a fair game.

There are no combinations of 5 through 8, 10 through 15, or 17 through 25 marbles for which the game is fair. We will prove this now.

One might notice that the numbers of red and green marbles in Table 4.1 are *triangular* numbers, that is, they are sums of consecutive positive integers $1 + 2 = 3$, $1 + 2 + 3 = 6$, $1 + 2 + 3 + 4 = 10$, and so on. The term *triangular* comes from the fact that these numbers can be shown as equilateral triangles:

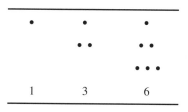

We also note that $1 + 2 + 3 + \cdots + k = k(k + 1)/2$. To see that this is so, note that the formula works for $k = 2$ and also note that if the formula is correct for the sum of k positive integers, then

$$1 + 2 + 3 + \cdots + k + (k + 1) = \frac{k(k + 1)}{2} + (k + 1) = \frac{(k + 2)(k + 1)}{2}$$

which is the formula for the sum of $k + 1$ positive integers. This proves the formula.

Let us see why R and G must be triangular numbers. For the game to be fair,

$$\binom{R}{2} + \binom{G}{2} = \frac{1}{2}\binom{R + G}{2}$$

or

$$2R(R - 1) + 2G(G - 1) = (R + G)(R + G - 1)$$

and this can easily be simplified to

$$R + G = (R - G)^2$$

This is one equation in two unknowns. But we also know that both R and G must be positive integers. So let $R - G = k$. Then, since $R + G = (R - G)^2$, it follows that $R + G = k^2$.

Solving these simultaneously gives $2R = k + k^2$ or $R = k(k+1)/2$ and so $G = k(k-1)/2$ showing that R and G are consecutive triangular numbers.

EXAMPLE 4.7   *Randomized Response*

We show here a geometric solution to the randomized response exploration in Chapter 2.

Those who do sample surveys are often interested in asking sensitive questions such as "Are you an illegal drug user?" or "Have you ever committed a serious crime?" Asking these questions directly would involve self-incrimination and would likely not produce honest answers, so it would be unlikely that we could determine the percentage of drug users or dangerous criminals with any degree of accuracy.

Here is a procedure for obtaining responses to sensitive survey questions that has proved to be quite accurate. Suppose we have two questions:

1. Is your Social Security number even?

2. Are you an illegal drug user?

The interviewer then asks the person being interviewed to flip a fair coin (and not show the result to the interviewer). If the coin comes up heads, he is to answer the first question; if the coin comes up tails, he is to answer the second question.

So if the person answers "Yes," we have no way of knowing whether his Social Security number is even or if he is a drug user. But it is possible, if we draw a picture of the situation, to estimate the proportion of drug users. Figure 4.9 should be very useful.

The square in the bottom left-hand corner has area $1/4$, representing those who when interviewed showed a head on the coin (probability $1/2$) and who then answered the first question "Yes" (also with probability $1/2$). The rectangle on the bottom right-hand side represents

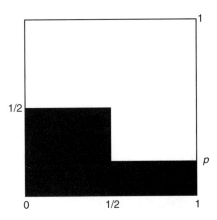

Figure 4.9

those who when interviewed showed a tail on the coin (probability $1/2$) and who answered the drug question "Yes." So the total area represents the proportion of those interviewed who responded "Yes."

Suppose that proportion of people answering "Yes" is, say, 0.30. Then , comparing areas, we have

$$1/4 + (1/2) \cdot p = 0.30$$

so

$$p = 2(0.30 - 1/4)$$

or

$$p = 0.10$$

So our estimate from this survey is that 10% of the population uses drugs.   ∎

This estimate is of course due to some sampling variability, namely, in the proportion showing heads on the coin. This should not differ much from $1/2$ if the sample is large, but could vary considerably from $1/2$ in a small sample.

It is useful to see how our estimate of p, the true proportion who should answer "Yes" to the sensitive question, varies with the proportion answering "Yes" in the survey, say p_s. We have that

$$1/4 + (1/2) \cdot p = p_s$$

so that

$$p = 2(p_s - 1/4)$$

A graph of this straight line is shown in Figure 4.10 where we assume that $1/4 \leq p_s \leq 3/4$ since if $p_s = 1/4$, $p = 0$, and if $p_s = 3/4$, then $p = 1$.

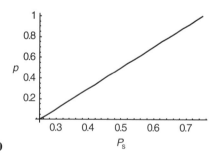

Figure 4.10

CONCLUSION

We have shown several examples in this chapter where problems in probability can be used in geometry. These are unusual examples for a geometry class and they may well motivate the student to draw graphs and draw conclusions from them. They are also examples of practical situations producing reasons to find areas and equations of straight lines, motivation that is often lacking in our classrooms.

EXPLORATIONS

Here are some interesting situations for classroom projects or group work.

1. Change the waiting times so that they are different for Joan and for Jim at the library.

2. In the foul shooting example, suppose we do not know the frequency with which C makes a foul shot, but we know the overall percentage of foul shots made by the team. How good a foul shooter is C?

3. Suppose in randomized response that we have the subject interviewed draw a ball from a sack of balls numbered from 1 to 100. If the number drawn (unknown to the interviewer) is less than or equal to 60, he is to answer the sensitive question; if the number is between 61 and 80, he is to simply say "Yes," otherwise, "No." What is the estimate of the "Yes" answers to the sensitive question?

4. Two numbers are chosen at random between 0 and 20.
   (a) What is the probability that their sum is less than 25?
   (b) What is the probability that the sum of their squares is less than 25?

5. In Exploration 4, draw pairs of random numbers with your computer or calculator and estimate the probability that the product of the numbers is less than 50.

Chapter 5

Random Variables and Discrete Probability Distributions—Uniform, Binomial, Hypergeometric, and Geometric Distributions

CHAPTER OBJECTIVES:

- to introduce random variables and probability distribution functions
- to discuss uniform, binomial, hypergeometric, and geometric probability distribution functions
- to discover some surprising results when random variables are added
- to encounter the "bell-shaped" curve for the first (but not the last!) time
- to use the binomial theorem with both positive and negative exponents.

INTRODUCTION

Suppose a hat contains slips of paper with the numbers 1 through 5 on them. A slip is drawn at random and the number on the slip observed. Since the result cannot be known in advance, the number is called a *random variable*. In general, a random variable is a variable that takes on values on the points of a sample space.

Random variables are generally denoted by capital letters, such as X, Y, Z, and so on. If we see the number 3 in the slip of paper experiment, we say that $X = 3$. It is important to distinguish between the variable itself, say X, and one of its values, usually denoted by small letters, say x.

A Probability and Statistics Companion, John J. Kinney
Copyright © 2009 by John Wiley & Sons, Inc.

Discrete random variables are those random variables that take on a finite, or perhaps a countably infinite, number of values so the associated sample space has a finite, or countably infinite, number of points. We discuss several discrete random variables in this chapter. Later, we will discuss several random variables that can take on an uncountably infinite number of values; these random variables are called *continuous* random variables.

If the random variable is discrete, we call the function giving the probability the random variable takes on any of its values, say $P(X = x)$, the *probability distribution function* of the random variable X. We will often abbreviate this as the PDF for the random variable X.

Random variables occur with different properties and characteristics; we will discuss some of these in this chapter. We begin with the probability distribution function suggested by the example of drawing a slip of paper from a hat.

DISCRETE UNIFORM DISTRIBUTION

The experiment of drawing one of five slips of paper from a hat at random suggests that the probability of observing any of the numbers 1 through 5 is $1/5$, that is,

$$P(X = x) = 1/5 \quad \text{for } x = 1, 2, 3, 4, 5$$

is the PDF for the random variable X. This is called a *discrete uniform* probability distribution function.

Not any function can serve as a probability distribution function. All discrete probability distribution functions have these properties:

If $f(x) = P(X = x)$ is the PDF for a random variable X, then

1) $f(x) \geq 0$

2) $\displaystyle\sum_{\text{all}x} f(x) = 1$

These properties arise from the fact that probabilities must be nonnegative and since some event must occur in the sample space, the sum of the probabilities over the entire sample space must be 1.

In general, the *discrete uniform probability distribution function* is defined as

$$f(x) = P(x = x) = 1/n \quad \text{for } x = 1, 2, ..., n$$

It is easy to verify that $f(x)$ satisfies the properties of a discrete probability distribution function.

Mean and Variance of a Discrete Random Variable

We pause now to introduce two numbers by which random variables can be characterized, the *mean* and the *variance*.

The *mean* or the *expected value* of a discrete random variable X is denoted by μ_x or $E(X)$ and is defined as

$$\mu_x = E(X) = \sum_{\text{all } x} x \cdot f(x)$$

This then is a weighted average of the values of X and the probabilities associated with its values. In our example, we find

$$\mu_x = E(X) = 1 \cdot \frac{1}{5} + 2 \cdot \frac{1}{5} + 3 \cdot \frac{1}{5} + 4 \cdot \frac{1}{5} + 5 \cdot \frac{1}{5} = 3$$

In general, we have discrete uniform distribution

$$\mu_x = E(X) = \sum_{\text{all } x} x \cdot f(x) = \sum_{x=1}^{n} x \cdot \frac{1}{n} = \frac{1}{n} \sum_{x=1}^{n} x = \frac{1}{n} \cdot \frac{n(n+1)}{2} = \frac{n+1}{2}$$

In our case where $n = 5$, we have $\mu_x = E(X) = (5+1)/2 = 3$, as before. Since the expected value is a sum, we have

$$E(X \pm Y) = E(X) \pm E(Y)$$

if X and Y are random variables defined on the same sample space. If the random variable X is a constant, say $X = c$, then

$$E(X) = \sum_{\text{all } x} x \cdot f(x) = c \sum_{\text{all } x} f(x) = c \cdot 1 = c$$

We now turn to another descriptive measure of a probability distribution, the *variance*. This is a measure of how variable the probability distribution is. To measure this, we might subtract the mean value from each of the values of X and find the expected value of the result. The thinking here is that values that depart markedly from the mean value show that the probability distribution is quite variable. Unfortunately, $E(X - \mu) = E(X) - E(\mu) = \mu - \mu = 0$ for any random variable. The problem here is that positive deviations from the mean exactly cancel out negative deviations, producing 0 for any random variable. So we square each of the deviations to prevent this and find the expected value of those deviations. We call this quantity the *variance* and denote it by

$$\sigma^2 = \text{Var}(X) = E(X - \mu)^2$$

This can be expanded as

$$\sigma^2 = \text{Var}(X) = E(X - \mu)^2 = E(X^2 - 2\mu X + \mu^2) = E(X^2) - 2\mu E(X) + \mu^2$$

or

$$\sigma^2 = E(X^2) - \mu^2$$

We will often use this form of σ^2 for computation. For the general discrete uniform distribution, we have

$$\sigma^2 = \sum_{x=1}^{n} x^2 \cdot \frac{1}{n} - \left(\frac{n+1}{2}\right)^2 = \frac{1}{n} \cdot \frac{n(n+1)(2n+1)}{6} - \frac{(n+1)^2}{4}$$

which is easily simplified to $\sigma^2 = (n^2 - 1)/12$. If $n = 5$, this becomes $\sigma^2 = 2$.
The positive square root of the variance, σ, is called the *standard deviation*.

Intervals, σ, and German Tanks

Probability distributions that contain extreme values (with respect to the mean) in general have larger standard deviations than distributions whose values are mostly close to the mean. This will be dealt later in this book when we consider *confidence intervals*.

For now, let us look at some uniform distributions and the percentage of those distributions contained in an interval centered at the mean. Since $\mu = (n + 1)/2$ and $\sigma = \sqrt{(n^2 - 1)/12}$ for the uniform distribution on $x = 1, 2, \ldots, n$, consider the interval $\mu \pm \sigma$ that in this case is the interval

$$\frac{n+1}{2} \pm \sqrt{\frac{n^2 - 1}{12}}$$

We have used the mean plus or minus one standard deviation here.

If $n = 10$, for example, this is the interval $(2.628, 8.372)$ that contains the values $3, 4, 5, 6, 7, 8$ or 60% of the distribution.

Suppose now that we add and subtract k standard deviations from the mean for the general uniform discrete distribution, then the length of this interval is $2k\sqrt{(n^2 - 1)/12}$ and the percentage of the distribution contained in that interval is

$$\frac{2k\sqrt{\dfrac{n^2 - 1}{12}}}{n} = 2k\sqrt{\frac{n^2 - 1}{12n^2}}$$

The factor $(n^2 - 1)/n^2$, however, rapidly approaches 1 as n increases. So the percentage of the distribution contained in the interval is approximately $2k/\sqrt{12} = k/\sqrt{3}$ for reasonably large values of n. If $k = 1$, this is about 57.7% and if $k = 1.5$, this is about 86.6%; k of course must be less than $\sqrt{3}$ or else the entire distribution is covered by the interval.

So the more standard deviations we add to the mean, the more of the distribution we cover, and this is true for probability distributions in general. This really seems to be a bit senseless however since we know the value of n and so it is simple to figure out what percentage of the distribution is contained in any particular interval.

We will return to confidence intervals later.

For now, what if we do not know the value of n? This is an entirely different matter. This problem actually arose during World War II. The Germans numbered much of the military material it put into the battlefield. They numbered parts of motorized vehicles and parts of aircraft and tanks. So, when tanks, for example, were captured, the Germans unwittingly gave out information about how many tanks were in the field!

If we assume the tanks are numbered $1, 2, \ldots, n$ (here is our uniform distribution!) and we have captured tanks numbered 7, 13, and 42, what is n?

This is not a probability problem but a *statistical* one since we want to *estimate* the value of n from a sample. We will have much more to say about statistical problems in later chapters and this one in particular. The reader may wish to think about this problem and make his or her own estimate. Notice that the estimate must exceed 42, but by how much? We will return to this problem later.

We now turn to another extremely useful situation, that of sums.

Sums

Suppose in the discrete uniform distribution with $n = 5$ (our slips of paper example) we draw two slips of paper. Suppose further that the first slip is replaced before the second is drawn so that the sampling for the first and second slips is done under exactly the same conditions. What happens if we add the two numbers found. Does the sum also have a uniform distribution?

We might think the answer to this is "yes" until we look at the sample space for the sum shown below.

Sample	Sum	Sample	Sum
1,1	2	3,4	7
1,2	3	3,5	8
1,3	4	4,1	5
1,4	5	4,2	6
1,5	6	4,3	7
2,1	3	4,4	8
2,2	4	4,5	9
2,3	5	5,1	6
2,4	6	5,2	7
2,5	7	5,3	8
3,1	4	5,4	9
3,2	5	5,5	10
3,3	6		

Now we realize that sums of 4, 5, or 6 are fairly likely. Here is the probability distribution of the sum:

X	2	3	4	5	6	7	8	9	10
$f(x)$	$\dfrac{1}{25}$	$\dfrac{2}{25}$	$\dfrac{3}{25}$	$\dfrac{4}{25}$	$\dfrac{5}{25}$	$\dfrac{4}{25}$	$\dfrac{3}{25}$	$\dfrac{2}{25}$	$\dfrac{1}{25}$

and the graph is as shown in Figure 5.1.

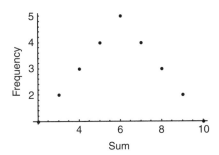

Figure 5.1

Even more surprising things occur when we increase the number of drawings, say to 5. Although the sample space now contains $5^5 = 3125$ points, enumerating these is a daunting task to say the least. Other techniques can be used however to produce the graph in Figure 5.2.

This gives a "bell-shaped" curve. As we will see, this is far from uncommon in probability theory; in fact, it is to be expected when sums are considered—and the sums can arise from virtually any distribution or combination of these distributions! We will discuss this further in the chapter on continuous probability distributions.

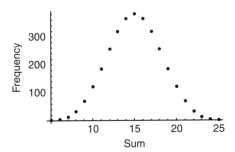

Figure 5.2

BINOMIAL PROBABILITY DISTRIBUTION

The next example of a discrete probability distribution is called the *binomial distribution*.

One of the most commonly occurring random variables is the one that takes one of two values each time the experiment is performed. Examples of this include tossing a coin, the result of which is a head or a tail; a newborn child is a female or male; a vaccination against the flu is successful or nonsuccessful. Examples of this situation are very common. We call these experiments *binomial* since, at each trial, the result is one of two outcomes, which, for convenience, are called success (S) or failure (F).

We will make two further assumptions: that the trials are independent and that the probabilities of success or failure remain constant from trial to trial. In fact, we let $P(S) = p$ and $P(F) = q = 1 - p$ for each trial. It is common to let the random variable X denote the number of successes in n independent trials of the experiment.

Let us consider a particular example. Suppose we inspect an item as it is coming off a production line. The item is good (G) or defective (D). If we inspect five items, the sample space then consists of all the possible sequences of five items, each G or D. The sample space then contains $2^5 = 32$ sample points. We also suppose as above that $P(G) = p$ and $P(D) = q = 1 - p$, and if we let X denote the number of good items, then we see that the possible values of X are 0, 1, 2, 3, 4, or 5. Now we must calculate the probabilities of each of these events.

If $X = 0$, then, unfortunately, none of the items are good so, using the independence of the events and the associated sample point,

$$P(X = 0) = P(DDDDD) = P(D) \cdot P(D) \cdot P(D) \cdot P(D) \cdot P(D) = q^5$$

How can $X = 1$? Then we must have exactly one good item and four defective items. But that event can occur in five different ways since the good item can occur at any one of the five trials. So

$$P(X = 1) = P(GDDDD \text{ or } DGDDD \text{ or } DDGDD \text{ or } DDDGD \text{ or } DDDDG)$$

$$= P(GDDDD) + P(DGDDD) + P(DDGDD)$$

$$+ P(DDDGD) + P(DDDDG)$$

$$= pq^4 + pq^4 + pq^4 + pq^4 + pq^4 = 5pq^4$$

Now $P(X = 2)$ is somewhat more complicated since two good items and three defective items can occur in a number of ways. Any particular order will have probability $q^3 p^2$ since the trials of the experiment are independent.

We also note that the number of orders in which there are exactly two good items must be $\binom{5}{2}$ or the number of ways in which we can select two positions for the two good items from five positions in total. We conclude that

$$P(X = 2) = \binom{5}{2} p^2 q^3 = 10p^2 q^3$$

In an entirely similar way, we find that

$$P(X = 3) = \binom{5}{3} p^3 q^2 = 10p^3 q^2$$

and

$$P(X = 4) = \binom{5}{4} p^4 q = 5p^4 q \text{ and } P(X = 5) = p^5$$

If we add all these probabilities together, we find

$$P(X = 0) + P(X = 1) + P(X = 2) + P(X = 3) + P(X = 4) + P(X = 5)$$

$$= q^5 + 5pq^4 + 10p^2 q^3 + 10p^3 q^2 + 5p^4 q + p^5 \quad \text{which we recognize as}$$

$$= (q + p)^5 = 1 \quad \text{since } q + p = 1$$

Note that the coefficients in the binomial expansion add up to 32, so all the points in the sample space have been used.

The occurrence of the binomial theorem here is one reason the probability distribution of X is called the *binomial probability distribution*.

The above situation can be generalized. Suppose now that we have n independent trials, that X denotes the number of successes, and that $P(S) = p$ and $P(F) = q = 1 - p$. We see that

$$P(X = x) = \binom{n}{x} p^x q^{n-x} \quad \text{for } x = 0, 1, 2, \dots, n$$

This is the probability distribution function for the binomial random variable in general.

We note that $P(X = x) \geq 0$ and $\sum_{x=0}^{n} P(X = x) = \sum_{x=0}^{n} \binom{n}{x} p^x q^{n-x} = (q + p)^n = 1$, so the properties of a discrete probability distribution function are satisfied.

Graphs of binomial distributions are interesting. We show some here where we have chosen $p = 0.3$ for various values of n (Figures 5.3, 5.4, and 5.5).

The graphs indicate that as n increases, the probability distributions become more "bell shaped" and strongly resemble what we will call, in Chapter 8, a continuous normal curve. This is in fact the case, although this fact will not be pursued here. One reason for not pursuing this is that exact calculations involving the

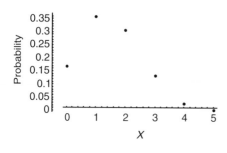

Figure 5.3 Binomial distribution, $n = 5$, $p = 0.3$.

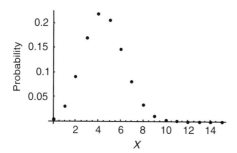

Figure 5.4 Binomial distribution, $n = 15$, $p = 0.3$.

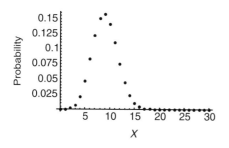

Figure 5.5 Binomial distribution $n = 30$, $p = 0.3$.

binomial distribution are possible using a statistical calculator or a computer algebra system, and so we do not need to approximate these probabilities with a normal curve, which we study in the chapter on continuous distributions. Here are some examples.

EXAMPLE 5.1 *A Production Line*

A production line has been producing good parts with probability 0.85. A sample of 20 parts is taken, and it is found that 4 of these are defective. Assuming a binomial model, is this a cause for concern?

Let X denote the number of good parts in a sample of 20. Assuming that the probability a good part is 0.85, we find the probability the sample has at most 16 good parts is

$$P(X \le 16) = \sum_{x=0}^{16} \binom{20}{x}(0.85)^x(0.15)^{20-x} = 0.352275$$

So this event is not unusual and one would probably conclude that the production line was behaving normally and that although the percentage of defective parts has increased, the sample is not a cause for concern. ■

EXAMPLE 5.2   *A Political Survey*

A sample survey of 100 voters is taken where actually 45% of the voters favor a certain candidate. What is the probability that the sample will contain between 40% and 60% of voters who favor the candidate?

We presume that a binomial model is appropriate. Note that the sample proportion of voters, say p_s, can be expressed in terms of the number of voters, say X, who favor the candidate. In fact, $p_s = X/100$, so

$$P(0.40 \le p_s \le 0.60) = P\left(0.40 \le \frac{X}{100} \le 0.60\right)$$

$$= P(40 \le X \le 60)$$

$$= \sum_{x=40}^{60} \binom{100}{x}(0.45)^x(0.55)^{100-x} = 0.864808$$

So if candidates in fact did not know the true proportion of voters willing to vote for them, the survey would be of little comfort since it indicates that they could either win or lose. Increasing the sample size will, of course, increase the probability that the sample proportion is within the range of 40–60%. ■

EXAMPLE 5.3   *Seed Germinations*

A biologist studying the germination rate of a certain type of seed finds that 90% of the seeds germinate. She has a box of 45 seeds.

The probability that all the seeds germinate, letting X denote the number of seeds germinating, is

$$P(X = 45) = \binom{45}{45}(0.90)^{45}(0.10)^0$$

$$= (0.90)^{45} = 0.0087280$$

So this is not a very probable event although the germination rate is fairly high.

The probability that at least 40 of the seeds germinate is

$$P(X \geq 40) = \sum_{x=40}^{45} \binom{45}{x} (0.90)^x (0.10)^{45-x}$$

$$= 0.70772$$

Now suppose that the seller of the seeds wishes to advertise a "guarantee" that at least k of the seeds germinates. What should k be if the seller wishes to disappoint at most 5% of the buyers of the seeds?

Here, we wish to determine k so that $P(X \geq k) = 0.05$, or $P(X \leq k) = 0.95$. This can be done only approximately.

Using a statistical calculator, we find the following in Table 5.1:

Table 5.1

k	$P(X \leq k)$
40	0.472862
41	0.671067
42	0.840957
43	0.947632
44	0.991272
45	1

It appears that one should fix the guarantee at 43 seeds. Since the variable is discrete, the cumulative probabilities increase in discrete jumps, so it is not possible to find a 5% error rate exactly. The table in fact indicates that this probability could be approximately 16%, 5%, or 1%, but no value in between these values. ∎

Mean and Variance of the Binomial Distribution

It will be shown in the next section that the following formulas apply to the binomial distribution with n trials and p the probability of success at any trial.

$$\mu = E(X) = \sum_{x=0}^{n} x \cdot \binom{n}{x} p^x (1-p)^{n-x} = np$$

$$\sigma^2 = Var(X) = E(X - \mu)^2 = \sum_{x=0}^{n} (x - \mu)^2 \cdot \binom{n}{x} p^x (1-p)^{n-x} = npq$$

In our above example, we calculate the mean value, $E(X) = \mu = 45 \cdot 0.90 = 40.5$ and variance, $Var(X) = npq = 45 \cdot 0.90 \cdot 0.10 = 4.05$, meaning that the standard deviation is $\sigma = \sqrt{4.05} = 2.01246$. We have seen that the standard deviation is a measure of the variation in the distribution. To illustrate this, we calculate

$$P(\mu - \sigma \leq X \leq \mu + \sigma) = P(40.5 - 2.01246 \leq X \leq 40.5 + 2.01246)$$

$$= P(38.48754 \leq X \leq 42.51246)$$

$$= \sum_{38}^{43} \binom{45}{x}(0.90)^x(0.10)^{45-x} = 0.871934.$$

Notice that we must round off some of the results since X can only assume integer values. We also find that

$$P(\mu - 2\sigma \leq X \leq \mu + 2\sigma) = P(40.5 - 2 \cdot 2.01246 \leq X \leq 40.5 + 2 \cdot 2.01246)$$

$$= P(36.47508 \leq X \leq 44.52492)$$

$$= \sum_{36}^{45} \binom{45}{x}(0.90)^x(0.10)^{45-x} = 0.987970.$$

We will return to these intervals in the chapter on continuous distributions.

Sums

In our study of the discrete uniform distribution, when we took sums of independent observations, graphs of those sums became "bell shaped", and we indicated that graphs of sums in general became shaped that way. Could it be then that binomial probabilities could be considered to be sums? The answer to this is "yes". The reason is as follows:

Consider a binomial process with n trials and probability of success at any particular trial, p. We define a random variable now for each one of the n trials as follows:

$$X_i = \begin{cases} 1 & \text{if the } i\text{th trial is a success} \\ 0 & \text{if the } i\text{th trial is a failure} \end{cases}$$

and X_i then is 1 only if the ith trial is a success; it follows that the sum of the X_i's counts the total number of successes in n trials. That is,

$$X_1 + X_2 + X_3 + \cdots + X_n = X$$

so the binomial random variable X is in fact a sum.

This explains the "bell shaped" curve we see when we graph the binomial distribution. The identity $X_1 + X_2 + X_3 + \cdots + X_n = X$ also provides an easy way to calculate the mean and variance of X. We find that $E(X_i) = 1 \cdot p + 0 \cdot q = p$ and since the expected value of a sum is the sum of the expected values,

$$E(X) = E(X_1 + X_2 + X_3 + \cdots + X_n) = E(X_1) + E(X_2) + E(X_3) + \cdots + E(X_n)$$
$$= p + p + p + \cdots + p = np$$

as we saw earlier.

We will show later that the variance of a sum of independent random variables is the sum of the variances and $\text{Var}(X_i) = E(X_i^2) - [E(X_i)]^2$, and $E(X_i^2) = p$,

$$\text{Var}(X_i) = p - p^2 = p(1 - p) = pq$$

it follows that

$$\text{Var}(X_i) = \text{Var}(X_1) + \text{Var}(X_2) + \text{Var}(X_3) + \cdots + \text{Var}(X_n)$$
$$= pq + pq + pq + \cdots + pq = npq$$

HYPERGEOMETRIC DISTRIBUTION

We now consider another very useful and frequently occurring discrete probability distribution.

The binomial distribution assumes that the probability of an event remains constant from trial to trial. This is not always an accurate assumption. We actually have encountered this situation when we studied acceptance sampling in the previous chapter; now we make the situation formal.

As an example, suppose that a small manufacturer has produced 11 machine parts in a day. Unknown to him, the lot contains three parts that are not acceptable (D), while the remaining parts are acceptable and can be sold (G). A sample of three parts is taken, and the sampling being done without replacement; that is, a selected part is not replaced so that it cannot be sampled again.

This means that after the first part is selected, no matter whether it is good or defective, the probability that the second part is good depends on the quality of the first part. So the binomial model does not apply.

We can, for example, find the probability that the sample of three contains two good and one unacceptable part. This event could occur in three ways.

$$P(2G, 1D) = P(GGD) + P(GDG) + P(DGG)$$
$$= \frac{8}{11} \cdot \frac{7}{10} \cdot \frac{3}{9} + \frac{8}{11} \cdot \frac{3}{10} \cdot \frac{7}{9} + \frac{3}{11} \cdot \frac{8}{10} \cdot \frac{7}{9} = 0.509$$

Notice that there are three ways for the event to occur, and each has the same probability. Since the order of the parts is irrelevant, we simply need to choose two of the good items and one of the defective items. This can be done in $\binom{8}{2} \cdot \binom{2}{1} = 56$ ways. So

$$P(2G, 1D) = \frac{\binom{8}{2} \cdot \binom{3}{1}}{\binom{11}{3}} = \frac{84}{165} = 0.509$$

as before. This is called a *hypergeometric probability distribution function.*

We can generalize this as follows. Suppose a manufactured lot contains D defective items and $N - D$ good items. Let X denote the number of unacceptable items in a sample of n items. This is

$$P(X = x) = \frac{\binom{D}{x} \cdot \binom{N - D}{n - x}}{\binom{N}{n}}, \quad x = 0, 1, \ldots, \text{Min}(n, D)$$

Since the sum covers all the possibilities

$$\sum_{x=0}^{\text{Min}(n,D)} \binom{D}{x} \cdot \binom{N - D}{n - x} = \binom{N}{n},$$

the probabilities sum to 1 as they should.

It can be shown that the mean value is $\mu_x = n \cdot (D/N)$, surprisingly like $n \cdot p$ in the binomial. The nonreplacement does not affect the mean value. It does effect the variance, however. The demonstration will not be given here, but

$$\text{Var}(X) = n \cdot \frac{D}{N} \cdot \left(1 - \frac{D}{N}\right) \cdot \frac{N - n}{N - 1}$$

is like the binomial *npq* except for the factor $(N - n)/(N - 1)$, which is often called a *finite population correction factor.*

To continue our example, we find the following values for the probability distribution function:

x	0	1	2	3
$P(X = x)$	$\dfrac{56}{165}$	$\dfrac{28}{55}$	$\dfrac{8}{55}$	$\dfrac{1}{165}$

Now, altering our sampling procedure from our previous discussion of acceptance sampling, suppose our sampling is destructive and we can only replace the defective items that occur in the sample. Then, for example, if we find one defective item in the sample, we sell 2/11 defective product. So the average defective product sold under this sampling plan is

$$\frac{56}{165} \cdot \frac{3}{11} + \frac{28}{55} \cdot \frac{2}{11} + \frac{8}{55} \cdot \frac{1}{11} + \frac{1}{165} \cdot \frac{0}{11} = 19.8\%$$

an improvement over the $3/11 = 27.3\%$ had we proceeded without the sampling plan.

The improvement is substantial, but not as dramatic as what we saw in our first encounter with acceptance sampling.

Other Properties of the Hypergeometric Distribution

Although the nonreplacement in sampling creates quite a different mathematical situation than that we encountered with the binomial distribution, it can be shown that the hypergeometric distribution approaches the binomial distribution as the population size increases.

It is also true that the graphs of the hypergeometric distribution show the same "bell-shaped" characteristic that we have encountered several times now, and it will be encountered again.

We end this chapter with another probability distribution that we have actually seen before, the geometric distribution. There are hundreds of other discrete probability distributions. Those considered here are a sample of these, although the sampling has been purposeful; we have discussed some of the most common distributions.

GEOMETRIC PROBABILITY DISTRIBUTION

In the binomial distribution, we have a fixed number of trials, and the random variable is the number of successes. In many situations, however, we wait for the first success, and the number of trials to achieve that success is the random variable.

In Examples 1.4 and 1.12, we discussed a sample space in which we sampled items emerging from a production line that can be characterized as good (G) or defective (D). We discussed a *waiting time* problem, namely, waiting for a defective item to occur. We presumed that the selections are independent and showed the following sample space:

$$S = \left\{ \begin{array}{c} D \\ GD \\ GGD \\ GGGD \\ \cdot \\ \cdot \\ \cdot \end{array} \right\}$$

Later, we showed that no matter the size of the probability an item was good or defective, the probability assigned to the entire sample space is 1.

Notice that in the binomial random variable, we have a fixed number of trials, say n, and a variable number of successes. In waiting time problems, we have a given number of successes (here 1); the number of trials to achieve those successes is the random variable.

Let us begin with the following waiting time problem. In taking a driver's test, suppose that the probability the test is passed is 0.8, the trials are independent, and the probability of passing the test remains constant. Let the random variable X denote the number of trials necessary up to and including when the test is passed. Then, applying the assumptions we have made and letting $q = 1 - p$, we find the following sample space (where T and F indicate respectively, that the test is passed and test has been failed), values of X, and probabilities in Table 5.2.

Table 5.2

Sample Point	X	Probability
T	1	$p = 0.8$
FT	2	$qp = (0.2)(0.8)$
FFT	3	$q^2 p = (0.2)^2(0.8)$
FFFT	4	$q^3 p = (0.2)^3(0.8)$
$\vdots$	$\vdots$	$\vdots$

We see that if the first success occurs at trial number x, then it must be preceded by exactly $x - 1$ failures, so we conclude that

$$P(X = x) = q^{x-1} \cdot p \quad \text{for } x = 1, 2, 3, \ldots$$

is the probability distribution function for the random variable X. This is called a *geometric* probability distribution function.

The probabilities are obviously positive and their sum is

$$S = p + qp + q^2 p + q^3 p + \cdots = 1$$

which we showed in Chapter 2 by multiplying the series by q and subtracting one series from another. The occurrence of a geometric series here explains the use of the word "geometric" in describing the probability distribution.

We also state here that $E[X] = 1/p$, a fact that will be proven in Chapter 7.

For example, if we toss a fair coin with $p = 1/2$, then the expected waiting time for the first head to occur is $1/(1/2) = 2$ tosses. The expected waiting time for our new driver to pass the driving test is $1/(8/10) = 1.25$ attempts.

We will consider the generalization of this problem in Chapter 7.

CONCLUSIONS

This has been a brief introduction to four of the most commonly occurring discrete random variables. There are many others that occur in practical problems, but those discussed here are the most important.

We will soon return to random variables, but only to random variables whose probability distributions are *continuous*. For now, we pause and consider an application of our work so far by considering seven-game series in sports. Then we will return to discrete probability distribution functions.

EXPLORATIONS

1. For a geometric random variable with parameters p and q, let r denote the probability that the first success occurs in no more than n trials.
   (a) Show that $r = 1 - q^n$.
   (b) Now let r vary and show a graph of r and n.

2. A sample of size 2 has been selected from the uniform distribution $1, 2, \cdots, n$, but it is not known whether $n = 5$ or $n = 6$. It is agreed that if the sum of the sample is 6 or greater, then it will be decided that $n = 6$. This decision rule is subject to two kinds of errors: we could decide that $n = 6$ while in reality $n = 5$, or we could decide that $n = 5$ while in reality $n = 6$. Find the probabilities of each of these errors.

3. A lot of 100 manufactured items contains an unknown number, k, of defective items. Items are selected from the lot and inspected, and the inspected items are not replaced before the next item is drawn. The second defective item is the fifth item drawn. What is k? (Try various values of k and select the value that makes the event most likely.)

4. A hypergeometric distribution has $N = 100$, with $D = 10$ special items. Samples of size 4 are selected. Find the probability distribution of the number of special items in the sample and then compare these probabilities with those from a binomial distribution with $p = 0.10$.

5. Flip a fair coin, or use a computer to select random numbers 0 or 1, and verify that the expected waiting time for a head to appear is 2.

Chapter 6

Seven-Game Series in Sports

CHAPTER OBJECTIVES:

- to consider seven-game play-off series in sports
- to discover when the winner of the series is in fact the better team
- to find the influence of winning the first game on winning the series
- to discuss the effect of extending the series beyond seven games.

INTRODUCTION

We pause now in our formal development of probability and statistics to concentrate on a particular application of the theory and ideas we have developed so far. Seven-game play-off series in sports such as basketball play-offs and the World Series present a waiting time problem. In this case, we wait until one team has won a given number of games. We analyze this problem in some depth.

SEVEN-GAME SERIES

In a seven-game series, the series ends when one team has won four games. This is another waiting time problem and gives us a finite sample space as opposed to the infinite sample spaces we have considered.

Let the teams be A and B with probabilities of wining an individual game as p and q, where $p + q = 1$. We also assume that p and q remain constant throughout the series and that the games are independent; that is, winning or losing a game has no effect on winning or losing the next game.

We look first at the sample space. The series can last four, five, six, or seven games. We show here the ways in which team A can win the series.

A Probability and Statistics Companion, John J. Kinney
Copyright © 2009 by John Wiley & Sons, Inc.

Four games	Five games	Six games		Seven games	
AAAA	BAAAA	BBAAAA	AABBAA	BBBAAAA	ABBBAAA
	ABAAA	BABAAA	AAABBA	BBABAAA	ABBABAA
	AABAA	BAABAA	ABABAA	BBAABAA	ABBAABA
	AAABA	BAAABA	ABAABA	BBAAABA	ABABBAA
		ABBAAA	AABABA	BABBAAA	ABABABA
				BABABAA	ABAABBA
				BABAABA	AABBBAA
				BAABBAA	AABBABA
				BAABABA	AABABBA
				BAAABBA	AAABBBA

To write out the points where B wins the series, interchange the letters A and B above. Note that the number of ways in which the series can be played in n games is easily counted. The last game must be won by A, say, so in the previous $n - 1$ games, A must win exactly three games and this can be done in $\binom{n-1}{3}$ ways. For example, there are $\binom{5}{3} = 10$ ways in which A can win the series in six games. So there are $\binom{4-1}{3} + \binom{5-1}{3} + \binom{6-1}{3} + \binom{7-1}{3} = 35$ ways for A to win the series and so 70 ways in which the series can be played.

These points are not equally likely, however, so we assign probabilities now to the sample points, where either team can win the series:

$$P(4 \text{ game series}) = p^4 + q^4$$
$$P(5 \text{ game series}) = 4p^4 q + 4q^4 p$$
$$P(6 \text{ game series}) = 10p^4 q^2 + 10q^4 p^2$$
$$P(7 \text{ game series}) = 20p^4 q^3 + 20q^4 p^3$$

These probabilities add up to 1 when the substitution $q = 1 - p$ is made. We also see that

$$P(A \text{ wins the series}) = p^4 + 4p^4 q + 10p^4 q^2 + 20p^4 q^3$$

and this can be simplified to

$$P(A \text{ wins the series}) = 35p^4 - 84p^5 + 70p^6 - 20p^7$$

by again making the substitution $q = 1 - p$.

This formula gives some interesting results. In Table 6.1, we show p, the probability that team A wins a single game, and P, the probability that team A wins the series.

Table 6.1

p	$P(A$ wins the series)
0.40	0.2898
0.45	0.3917
0.46	0.4131
0.47	0.4346
0.48	0.4563
0.49	0.4781
0.50	0.5000
0.51	0.5219
0.52	0.5437
0.53	0.5654
0.54	0.5869
0.55	0.6083
0.60	0.7102
0.70	0.8740
0.80	0.9667

It can be seen, if the teams are fairly evenly matched, that the probability of winning the series does not differ much from the probability of winning a single game!

The series is then not very discriminatory in the sense that the winner of the series is not necessarily the stronger team. The graph in Figure 6.1 shows the probability of winning the series and the probability of winning a single game. The maximum difference occurs when the probability of winning a single game is 0.739613 or 0.260387. The difference is shown in Figure 6.1.

What is the expected length of the series? To find this, we calculate

$$A = 4\left[p^4 + q^4\right] + 5\left[4p^4q + 4q^4p\right] + 6\left[10p^4q^2 + 10q^4p^2\right]$$
$$+7\left[20p^4q^3 + 20q^4p^3\right]$$

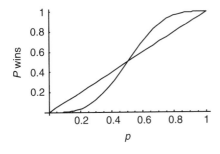

Figure 6.1   Probability of winning the series and the probability of winning a single game.

This can be simplified to

$$A = 4\left[1 + p + p^2 + p^3 - 13p^4 + 15p^5 - 5p^6\right]$$

A graph of A as a function of p is shown in Figure 6.2.

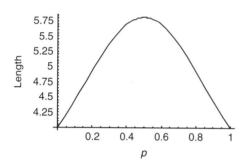

Figure 6.2   Expected length of the series.

The graph shows that in the range $0.45 \leq p \leq 0.55$, the average length of the series is almost always about 5.8 games.

WINNING THE FIRST GAME

Since the probability of winning the series is not much different from that of winning a single game, we consider the probability that the winner of the first game wins the series. From the sample space, we see that

$$P(\text{winner of the first game wins the series}) = p^3 + 3p^3q + 6p^3q^2 + 10p^3q^3$$

and this can be written as a function of p as

$$P(\text{winner of the first game wins the series}) = p^3(20 - 45p + 36p^2 - 10p^3)$$

A graph of A as a function of p is shown in Figure 6.3.

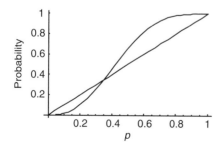

Figure 6.3   Probability that the winner of the first game wins the series.

The straight line gives the probability that an individual game is won. The other graph shows the probability that the winner of the first game wins the series. The graphs intersect at $p = 0.347129$; so, if p is greater than this, then the winner of the first game is more likely to win the series.

How Long Should the Series Last?

We have found that the winner of a seven-game series is not necessarily the better team. In fact, if the teams are about evenly matched, the probability that the weaker team wins the series is about equal to the probability that the team wins an individual game.

Perhaps a solution to this is to increase the length of the series so that the probability that the winner of the series is in fact the stronger team is increased.

How long, then, should the series last?

We presume the series is over when, in a series of n games, one team has won $(n + 1)/2$ games.

If $n = 7$, then the winner must win four games.

If $n = 9$, then the winner must win five games.

Now we find a formula for the probability that a given team wins a series of n games. Call the winner of the series A and suppose the probability that A wins an individual game is p and the probability that the loser of the series wins an individual game is q. If A wins the series, then A has won $(n + 1)/2$ games and has won $(n + 1)/2 - 1 = (n - 1)2$ games previously to winning the last game. During this time, the loser of the series has won x games (where x can be as low as 0 and at most $(n - 1)/2$ games). So there are $(n - 1)/2 + x = (n + 2x - 1)/2$ games played before the final game. It follows that

$$P(A \text{ wins the series}) = p^{(n+1)/2} \sum_{x=0}^{(n-1)/2} \binom{(n + 2x - 1)/2}{x} q^x$$

If $n = 7$, this becomes

$$P(A \text{ wins the series}) = p^4 \left[1 + \binom{4}{1}q + \binom{5}{2}q^2 + \binom{6}{3}q^3 \right]$$

and if $n = 9$, this becomes

$$P(A \text{ wins the series}) = p^5 \left[1 + \binom{5}{1}q + \binom{6}{2}q^2 + \binom{7}{3}q^3 + \binom{8}{4}q^4 \right]$$

Increasing the number of games to nine changes slightly the probability that A wins the series, as shown in Table 6.2.

Table 6.2 Probability of Winning the Series

p	Seven-game series	Nine-game series
0.45	0.391712	0.378579
0.46	0.413058	0.402398
0.47	0.434611	0.426525
0.48	0.456320	0.450886
0.49	0.478134	0.475404
0.50	0.500000	0.500000
0.51	0.521866	0.549114
0.52	0.543680	0.573475
0.53	0.565389	0.573475
0.54	0.586942	0.597602
0.55	0.608288	0.621421

The difference between the two series can be seen in Figure 6.4.

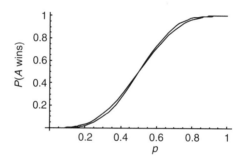

Figure 6.4

The steeper curve is that for a nine-game series.

Now suppose we want $P(A$ wins the series$)$ to be some high probability, say 0.95. This requires that the graph, such as one given in Figure 6.4, should pass through the point $(p, 0.95)$ for some value of n. Suppose again that p is the probability that A wins an individual game and that $p > 0.5$. How many games should be played?

To solve this in theory, put $P(A$ wins the series$) = 0.95$ and solve the resulting equation for n. The equation to be solved is then

$$p^{(n+1)/2} \sum_{x=0}^{(n-1)/2} \binom{(n + 2x - 1)/2}{x} q^x = 0.95$$

where we know p and q.

This cannot be done exactly. So, we defined a function $h[p, n]$ as follows:

$$h[p, n] = p^{(n+1)/2} \sum_{x=0}^{(n-1)/2} \binom{(n + 2x - 1)/2}{x} q^x$$

and then experimented with values of p and n until a probability of about 0.95 was achieved. The results are shown in Table 6.3.

Table 6.3

p	Number of games (n)
0.60	63
0.59	79
0.58	99
0.57	131
0.56	177
0.55	257
0.54	401
0.53	711
0.52	1601
0.51	4000*

It is apparent that the length of the series becomes far too great, even if $p = 0.60$, so that the teams are really quite unevenly matched. As the teams approach parity, the number of games required grows quite rapidly. In the case where $p = 0.51$, the number of games exceeds 4000! In fact $h[0.51, 4000] = 0.90$, so even with 4000 games (which is difficult to imagine!) we have not yet achieved a probability of 0.95! This should be overly convincing that waiting until a team has won $(n + 1)/2$ games is not a sensible plan for deciding which team is stronger. We will not pursue it here, but tennis turns out to be a fairly good predictor of who the better player is; this is because the winner of a game must win at least two points in a row and the winner of a set must win at least two games in a row.

CONCLUSIONS

We have studied seven-game series in sports and have concluded that the first team to win four games is by no means necessarily the stronger team. The winner of the first game has a decided advantage in winning the series.

Extending the number of games so that the probability that the winner of a lengthy series is most probably the better team is quite unfeasible.

EXPLORATIONS

1. Verify that the probabilities given for the seven-game series where p is the probability that a given team wins a game and $q = 1 - p$ is the probability that a given team loses a game add up to 1.

2. Compare the actual record of the lengths of the World Series that have been played to date with the probabilities of the lengths assuming the teams to be

evenly matched. Does the actual record suggest that the teams are not evenly matched?

3. Compare the actual expected length of the series with the theoretical expected length. Do the data suggest that the teams are not evenly matched?

4. Find data showing the winner of the first game in a seven-game series and determine how often that team wins the series.

Chapter 7

Waiting Time Problems

CHAPTER OBJECTIVES:

- to develop the negative binomial probability distribution
- to apply the negative binomial distribution to a quality control problem
- to show some practical applications of geometric series
- to show how to sum some series which are not geometric (without calculus)
- to encounter the Fibonacci sequence when tossing a coin
- to show an unfair game with a fair coin
- to introduce the negative hypergeometric distribution
- to consider an (apparent) logical contradiction.

$\mathbf{W}$e now turn our attention to some problems usually not considered in an introductory course in probability and statistics, namely, *waiting time* problems. As we have seen in problems involving the binomial probability distribution, we consider a fixed number of trials and calculate the probability of a given number of "successes." Now we consider problems where we wait for a success, or a given number of successes, or some pattern of successes and failures.

WAITING FOR THE FIRST SUCCESS

Recall, from Chapter 5, that in a binomial situation we have an experiment with one of the two outcomes, which, for lack of better terms, we call "success" and "failure." We must also have a constant probability of success, say p, and consequently a constant probability of failure, say q where of course $p + q = 1$. It is also necessary for the experiments, or trials, to be independent. In this situation, it is common to define a random variable, X, denoting the number of successes in n independent trials. We

A Probability and Statistics Companion, John J. Kinney
Copyright © 2009 by John Wiley & Sons, Inc.

have seen that the probability distribution function is

$$P(X = x) = \binom{n}{x} p^x q^{n-x}, x = 1, 2, \ldots, n$$

and we have verified that the resulting probabilities add up to 1.

Tossing a two-sided loaded coin is a perfect binomial model.

Now, however, we assume the binomial presumptions, but we do not have a fixed number of trials, rather we fix the number of successes and then the number of trials becomes the random variable.

Let us begin by waiting for the first success, as we did in Chapter 5.

The sample space and associated probabilities are shown in Table 7.1. S denotes a success and F denotes a failure.

Table 7.1

Sample point	Probability	Number of trials
S	p	1
FS	qp	2
FFS	$q^2 p$	3
$FFFS$	$q^3 p$	4
$\vdots$	$\vdots$	

Now, again using the symbol X, which now denotes the number of trials necessary to achieve the first success, it is apparent that since we must have $x - 1$ failures followed by a success

$$P(X = x) = q^{x-1} p, x = 1, 2, 3, \ldots$$

where the values for X now have no bound. We called this a *geometric* probability distribution.

We have shown that the probabilities sum to 1 so we have established that we have a true probability distribution function. Now we consider a very specific example and after that we will generalize this problem.

THE MYTHICAL ISLAND

Here is an example of our waiting time problem. On a mythical island, couples are allowed to have children until a male child is born. What effect, if any, does this have on the male:female ratio on the island? The answer may be surprising.

Suppose that the probabilities of a male birth or a female birth are each 1/2 (which is not the case in actuality) and that the births are independent of each other.

The sample space now is shown in Table 7.2.

We know that $S = p + qp + q^2 p + q^3 p + \cdots = 1$ and here $p = q = 1/2$ so we have a probability distribution. We now want to know the expected number of males

Table 7.2

Sample point	Probability
M	$\dfrac{1}{2}$
FM	$\dfrac{1}{2} \cdot \dfrac{1}{2} = \dfrac{1}{4}$
FFM	$\dfrac{1}{2} \cdot \dfrac{1}{2} \cdot \dfrac{1}{2} = \dfrac{1}{8}$
$FFFM$	$\dfrac{1}{2} \cdot \dfrac{1}{2} \cdot \dfrac{1}{2} \cdot \dfrac{1}{2} = \dfrac{1}{16}$
$\vdots$	$\vdots$

in a family. To find this, let

$$A_M = 1 \cdot \frac{1}{2} + 1 \cdot \frac{1}{4} + 1 \cdot \frac{1}{8} + 1 \cdot \frac{1}{16} + \cdots$$

then

$$\frac{1}{2} A_M = 1 \cdot \frac{1}{4} + 1 \cdot \frac{1}{8} + 1 \cdot \frac{1}{16} + \cdots$$

and subtracting the second series from the first series,

$$\frac{1}{2} A_M = \frac{1}{2}$$

so $A_M = 1$ Now the average family size is $A = 1 \cdot \frac{1}{2} + 2 \cdot \frac{1}{4} + 3 \cdot \frac{1}{8} + 4 \cdot \frac{1}{16} + \cdots$ and so

$$\frac{1}{2} A = 1 \cdot \frac{1}{4} + 2 \cdot \frac{1}{8} + 3 \cdot \frac{1}{16} + 4 \cdot \frac{1}{32} + \cdots$$

Subtracting these series gives

$$A - \frac{1}{2} A = 1 \cdot \frac{1}{2} + 1 \cdot \frac{1}{4} + 1 \cdot \frac{1}{8} + 1 \cdot \frac{1}{16} + \cdots = 1, \text{ so } A = 2.$$

Since the average number of male children in a family is 1, so is the average number of females, giving the male:female ratio as 1:1, just as it would be if the restrictive rule did not apply!

We now generalize the waiting time problem beyond the first success.

WAITING FOR THE SECOND SUCCESS

The sample space and associated probabilities when we wait for the second success are shown in Table 7.3. Note that we are not necessarily waiting for two successes in a row. We will consider that problem later in this chapter.

It is clear that among the first $x - 1$ trials, we must have exactly one success.

We conclude that if the second success occurs at the xth trial, then the first $x - 1$ trials must contain exactly one success (and is the usual binomial process), and this is followed by the second success when the experiment ends.

Table 7.3

Sample point	Probability	Number of trials
SS	p^2	2
FSS	qp^2	3
SFS	qp^2	3
FFSS	q^2p^2	4
FSFS	q^2p^2	4
SFFS	q^2p^2	4
$\vdots$	$\vdots$	$\vdots$

We conclude that

$$P(X = x) = \binom{x-1}{1} pq^{x-2} \cdot p$$

or

$$P(X = x) = \binom{x-1}{1} p^2 q^{x-2}, x = 2, 3, 4, \cdots$$

Again we must check that we have assigned a probability of 1 to the entire sample space.

Adding up the probabilities we see that

$$\sum_{x=2}^{\infty} (X = x) = p^2 \sum_{x=2}^{\infty} \binom{x-1}{1} q^{x-2}$$

Call the summation T so

$$T = \sum_{x=2}^{\infty} \binom{x-1}{1} q^{x-2} = \sum_{x=2}^{\infty} (x-1)q^{x-2} = 1 + 2q + 3q^2 + 4q^3 + 5q^4 + \cdots$$

Now $qT = q + 2q^2 + 3q^3 + 4q^4 + \cdots$ and subtracting qT from T it follows that $(1-q)T = 1 + q + q^2 + q^3 + \cdots$ which is a geometric series. So $(1-q)T = 1/(1-q)$ and so $T = 1/(1-q)^2 = 1/p^2$ and since

$$\sum_{x=2}^{\infty} P(X = x) = p^2 \sum_{x=2}^{\infty} \binom{x-1}{1} q^{x-2}$$

it follows that

$$\sum_{x=2}^{\infty} P(X = x) = p^2 \cdot \frac{1}{p^2} = 1$$

We used the process of multiplying the series $T = 1 + q + 2q^2 + 3q^3 + 4q^4 + \cdots$ by q in order to sum the series because this process will occur again. We could also have noted that $T = 1/(1-q)^2 = (1-q)^{-2} = 1/p^2$ by the binomial expansion with a negative exponent. We will generalize this in the next section and see that our last two examples are special cases of what we will call the *negative binomial* distribution.

WAITING FOR THE *r*th SUCCESS

We now generalize the two special cases we have done and consider waiting for the *r*th success where $r = 1, 2, 3, \ldots$. Again the random variable X denotes the waiting time or the total number of trials necessary to achieve the *r*th success.

It is clear that if the xth trial is the rth success, then the previous $x - 1$ trials must contain exactly $r - 1$ successes by a binomial process and in addition the rth success occurs on the xth trial. So

$$\sum_{x=r}^{\infty} P(X = x) = \sum_{x=r}^{\infty} \binom{x-1}{r-1} p^{r-1} q^{x-r} \cdot p = p^r \sum_{x=r}^{\infty} \binom{x-1}{r-1} q^{x-r}$$

As in the special cases, consider

$$T = \sum_{x=r}^{\infty} \binom{x-1}{r-1} q^{x-r} = 1 + \binom{r}{r-1} q + \binom{r+1}{r-1} q^2 + \binom{r+2}{r-1} q^3 + \cdots$$

But this is the expansion of $(1 - q)^{-r}$, so

$$\sum_{x=r}^{\infty} P(X = x) = p^r \sum_{x=r}^{\infty} \binom{x-1}{r-1} q^{x-r} = p^r (1-q)^{-r} = p^r p^{-r} = 1$$

so our assignment of probabilities produces a probability distribution function. The function is known as the *negative binomial* distribution due to the occurrence of a binomial expansion with a negative exponent and is defined the way we have above as

$$P(X = x) = \binom{x-1}{r-1} p^r q^{x-r}, x = r, r+1, r+2, \ldots$$

When $r = 1$ we wait for the first success and the probability distribution becomes $P(X = x) = pq^{x-1}, x = 1, 2, 3, \ldots$ as we saw above; if $r = 2$, the probability distribution becomes $P(X = x) = (x - 1)p^2 q^{x-2}, x = 2, 3, 4, \ldots$ again as we found above.

MEAN OF THE NEGATIVE BINOMIAL

The expected value of the negative binomial random variable is easy to find.

$$E[X] = \sum_{x=r}^{\infty} x \cdot \binom{x-1}{r-1} p^r q^{x-r} = p^r \cdot r \sum_{x=r}^{\infty} \binom{x}{r} q^{x-r}$$

$$= p^r \cdot r \left[1 + \binom{r+1}{1} q + \binom{r+2}{2} q^2 + \binom{r+3}{3} q^3 + \cdots \right]$$

The quantity in the square brackets is the expansion of $(1 - q)^{-(r+1)}$ and so

$$E[X] = p^r \cdot r \cdot (1 - q)^{-(r+1)} = \frac{r}{p}$$

If p is fixed, this is a linear function of r as might be expected. If we wait for the first head in tossing a fair coin, $r = 1$ and $p = 1/2$ so our average waiting time is two tosses.

COLLECTING CEREAL BOX PRIZES

A brand of cereal promises one of six prizes in a box of cereal. On average, how many boxes must a person buy in order to collect all the prizes?

Here r above remains at 1 but the value of p changes as we collect the coupons. The first box yields the first prize, but then the average waiting time to find the next prize is $1/(5/6)$, the average waiting time for the next prize is $1/(4/6)$, and so on, giving the total number of boxes bought on average to be $1 + 1/(5/6) + 1/(4/6) + 1/(3/6) + 1/(2/6) + 1/(1/6) = 14.7$ boxes. Note that the waiting time increases as the number of prizes collected increases.

HEADS BEFORE TAILS

Here is another game with a coin, this time a loaded one.

Let p be the probability a head occurs when the coin is tossed. A running count of the heads and tails is kept; we want the probability that the heads count reaches three before the tails count reaches four. Let us call this event "3 heads before 4 tails".

If the event is to occur, we must throw the third head on the last trial and this must be preceded by at most three tails. So if we let x denote the number of tails then the random variable x must be 0, 1, 2, or 3. So the tails must occur in the first $2 + x$ trials (we need two heads and x tails) and of course we must have three heads in the final result.

The probability of this is then

$$P(3 \text{ heads before 4 tails}) = p^3 \sum_{x=0}^{3} (1 - p)^x \binom{2 + x}{2}$$

$$= p^3[1 + 3q + 6q^2 + 10q^3] = p^3[6(1 - p)^2 - 3p + 10(1 - p)^3 + 4]$$

$$= 36p^5 - 45p^4 - 10p^6 + 20p^3$$

A graph of this function is shown in Figure 7.1 for various values of p.

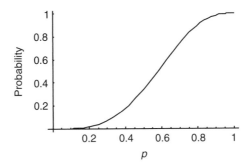

Figure 7.1

Let us generalize the problem so that we want the probability that a heads occur before b tails.

The last toss must be a head. Then, of the preceding tosses, exactly $a - 1$ must be heads and x must be tails, where x is at most $b - 1$, so

$$P(a \text{ heads before b tails}) = \sum_{x=0}^{b-1} p^a q^x \binom{a - 1 + x}{x}$$

Now let us fix the number of tails, say let $b = 5$. (Note that a was fixed above at 3). So

$$P(a \text{ heads before 5 tails}) = \sum_{x=0}^{4} p^a q^x \binom{a - 1 + x}{x}$$

A graph of this function is in Figure 7.2 where we have taken $p = 0.6$.

Finally, we show in Figure 7.3 a graph of the surface when both a and p are allowed to vary.

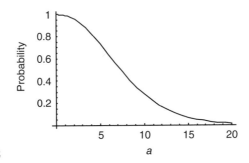

Figure 7.2

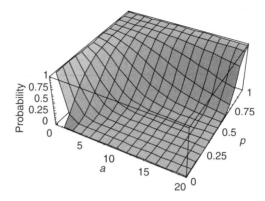

Figure 7.3

WAITING FOR PATTERNS

We have considered the problem of waiting for the rth head in coin tossing; now we turn to some unusual problems involving waiting for more general patterns in binomial trials. We will encounter some interesting mathematical consequences and we will show an unfair game with a fair coin.

We begin with a waiting time problem that involves waiting for two successes in a row. Note that this differs from waiting for the second success that is a negative binomial random variable. Let us begin with the points in the sample space where we have grouped the sample points by the number of experiments necessary. These are shown in Table 7.4.

Table 7.4

Sample point	Number of points
HH	1
THH	1
TTHH *HTHH*	2
TTTHH *THTHH* *HTTHH*	3
TTTTHH *TTHTHH* *THTTHH* *HTTTHH* *HTHTHH*	5
$\vdots$	$\vdots$

We see that the number of points is $1, 1, 2, 3, 5, \ldots$. The Fibonacci sequence begins with $1, 1$; subsequent terms are found by adding the two immediately preceding terms. Can it be that the number of points follows the Fibonacci sequence? Of course, we cannot conclude that just because the pattern holds in the first few cases.

But the Fibonacci pattern does hold here! Here is why: if two heads in a row occur on the nth toss, then either the sequence begins with T followed by HH in $n - 1$ tosses or the sequence begins with HT (to avoid the pattern HH) followed by HH in $n - 2$ tosses. So the number of points in the sample space is found by writing T before every point giving HH in $n - 1$ tosses and writing HT before every point giving HH in $n - 2$ tosses. So the total number of points in the sample space for the occurrence of HH in n tosses is the sum of the number of points for which HH occurs in $n - 1$ tosses and the number of points in which HH occurs in $n - 2$ tosses, the Fibonacci sequence.

Here, for example, using the points in the previous table, are the points for which HH occurs for the first time at the seventh toss:

$$T|TTTTHH$$
$$T|TTHTHH$$
$$T|THTTHH$$
$$T|HTTTHH$$
$$T|HTHTHH$$

$$HT|TTTHH$$
$$HT|THTHH$$
$$HT|HTTHH$$

The assignment of probabilities to the sample points is easy, but the pattern they follow is difficult and cannot be simply stated.

EXPECTED WAITING TIME FOR HH

To calculate this expectation, let a_n denote the probability that the event HH occurs at the nth trial. Then, using the argument leading to the Fibonacci series above, it follows that

$$a_n = qa_{n-1} + qpa_{n-2} \text{ for } n > 2 \text{ and we take } a_1 = 0 \text{ and } a_2 = p^2.$$

This formula is a *recursion* and we will study this sort of formula in Chapter 16.

Now multiply this recursion through by n and sum this result from $n = 3$ to infinity to find

$$\sum_{n=3}^{\infty} na_n = q \sum_{n=3}^{\infty} na_{n-1} + qp \sum_{n=3}^{\infty} na_{n-2}$$

which we can also write as

$$\sum_{n=3}^{\infty} na_n = q \sum_{n=3}^{\infty} [(n-1)+1]a_{n-1} + qp \sum_{n=3}^{\infty} [(n-2)+2]a_{n-2}$$

or

$$\sum_{n=3}^{\infty} na_n = q \sum_{n=3}^{\infty} (n-1)a_{n-1} + q \sum_{n=3}^{\infty} a_{n-1} + qp \sum_{n=3}^{\infty} (n-2)a_{n-2} + 2qp \sum_{n=3}^{\infty} a_{n-2}$$

This simplifies, since

$$\sum_{n=1}^{\infty} na_n = E[N] \text{ and } \sum_{n=1}^{\infty} a_n = 1,$$

to

$$E[N] - 2a_2 = qE[N] + q + qpE[N] + 2qp$$ or

$(1 - q - qp)E[N] = 2p^2 + q + 2qp = 1 + p$ so

$E[N] = (1+p)/(1-q-qp) = (1+p)/p^2$. This can also be written as $E[N] = 1/p^2 + 1/p$.

It might be thought that this expectation would be just $1/p^2$, but that is not quite the case. For a fair coin, this expectation is six tosses.

It is fairly easy to see that if we want for *HHH* then we get a "super" Fibonacci sequence in which we start with 1, 1, 1 and obtain subsequent terms by adding the previous three terms in the sequence.

As a second example, let us consider waiting for the pattern *TH*. Table 7.5 shows some of the sample points.

Now the number of points in the sample space is simple. Suppose *TH* occurs on the *n*th toss. The sample point then begins with 0, 1, 2, 3, ..., $n-2$ *H*'s. So there are $n-1$ points for which *TH* occurs on the *n*th toss.

Table 7.5

Sample point
TH
TTH
HTH
TTTH
HTTH
HHTH
TTTTH
HTTTH
HHTTH
HHHTH
⋮

This observation makes the sample space fairly easy to write out and it also makes the calculation of probabilities fairly simple. Note that the probability of any sample point has the factor qp for the sequence TH at the nth toss.

If the sample point begins with no heads, then its probability has a factor of q^{n-2}.

If the sample point begins with H, then its probability has a factor of pq^{n-3}.

If the sample point begins with HH, then its probability has a factor of p^2q^{n-4}.

This pattern continues until we come to the sample point with $n-2$ H's followed by TH. The probability of this point has a factor of p^{n-2}.

Consider the case for $n = 5$ shown in the sample space above. The probabilities of the points add to

$$qp(q^3 + pq^2 + p^2q + p^3)$$

This can be recognized as $qp(q^4 - p^4)/(q - p)$. This pattern continues, and by letting X denote the total number of tosses necessary, we find that

$$P(X = n) = \frac{qp(q^{n-1} - p^{n-1})}{q - p} \qquad \text{for } n = 2, 3, 4, \ldots$$

The formula would not work for $q = p = 1/2$.

In that case the sample points are equally likely, each having probability $(1/2)^n$ and, as we have seen, there are $n - 1$ of them. It follows in the case where $q = p = 1/2$ that

$$P(X = n) = \frac{n - 1}{2^n} \qquad \text{for } n = 2, 3, 4, \ldots$$

EXPECTED WAITING TIME FOR TH

Using the formula above,

$$E[N] = \sum_{n=2}^{\infty} n \cdot \frac{qp(q^{n-1} - p^{n-1})}{q - p}$$

To calculate this sum, consider

$$S = \sum_{n=2}^{\infty} n \cdot q^{n-1} = 2q + 3q^2 + 4q^3 + \cdots$$

Now $S + 1 = 1 + 2q + 3q^2 + 4q^3 + \cdots$ and we have seen that the left-hand side of this equation is $1/p^2$. So $S = 1/p^2 - 1$.

$$E[N] = \frac{qp}{q-p}\left[\frac{1}{p^2} - 1 - \left(\frac{1}{q^2} - 1\right)\right] = \frac{1}{qp}.$$

The formula above applies only if $p \neq q$. In the case $p = q$, we first consider

$$\frac{P(X = n+1)}{P(X = n)} = \frac{n}{2(n-1)}$$

So

$$\sum_{n=2}^{\infty} 2(n-1)P(X = n+1) = \sum_{n=2}^{\infty} n P(X = n).$$

We can write this as

$$2\sum_{n=2}^{\infty}[(n+1) - 2]P(X = n+1) = \sum_{n=2}^{\infty} n P(X = n)$$

and from this it follows that $2E[N] - 4 = E[N]$ so $E[N] = 4$.

This may be a surprise. We found that the average waiting time for HH with a fair coin is six. (By symmetry, the average waiting time for TT is six tosses and the average waiting time for HT is four tosses.)

There is apparently no intuitive reason for this to be so, but these results can easily be verified by simulation to provide some evidence that they are correct.

We will simply state the average waiting times for each of the patterns of length 3 with a fair coin in Table 7.6.

Table 7.6

Pattern	Average waiting time
HHH	14
THH	8
HTH	10
TTH	8
THT	10
HTT	8
HHT	8
TTT	14

We continue this chapter with an intriguing game.

AN UNFAIR GAME WITH A FAIR COIN

Here is a game with a fair coin. In fact, it is an unfair game with a fair coin!

Consider the patterns *HH*, *TH*, *HT*, and *TT* for two tosses of a fair coin. If you choose one of these patterns, I will choose another and then we toss a fair coin until one of these patterns occurs. The winner is the person who chose the first-occurring pattern. For example, if you choose *HH*, I will choose *TH*. If we see the sequence *HTTTTH*, then I win since the pattern *TH* occurred before the pattern *HH*. My probability of beating you is 3/4! Here is why.

If the first two tosses are *HH*, you win.

If the first two tosses are *TH*, I win.

If the first two tosses are *HT* then this can be followed by any number of *T*'s, but eventually *H* will occur and I win.

If the first two tosses are *TT* then this can be followed by any number of *T*'s but eventually *H* will occur and I will win.

If you are to win, you must toss *HH* on the first two tosses; this is the only way you can win and it has probability 1/4. If a *T* is tossed at any time, I will win, so my probability of winning is 3/4.

The fact that the patterns *HH*, *TH*, *HT*, and *TT* are equally likely for a fair coin may be observed by a game player who may think that any choice is equally good is irrelevant; the choice of pattern is crucial, as is the fact that he is allowed to make the first choice.

If you choose *TT*, then the only way you can win is by tossing two tails on the first two tosses. I will choose *HT* and I will win 3/4 of the time. A sensible choice for you is either *TH* or *HT*, and then my probability of beating you is only 1/2.

If we consider patterns with three tosses, as shown in the table above, my *minimum* probability of beating you is 2/3! (And if you do not choose well, I can increase that probability to 3/4 or 7/8!). This can be tried by simulation.

It is puzzling to note that no matter what pattern you choose, I will probably beat you. This means that if we play the game twice, and I win in the first game, then you can choose the pattern I chose on the first game, and I can still probably beat you.

Probabilities then are not transitive, so if pattern A beats pattern B and pattern B beats pattern C, then it does not follow that pattern A will necessarily beat pattern C.

THREE TOSSES

This apparent paradox, that probabilities are not transitive, continues with patterns involving patterns of length 3. We showed above the average waiting times for each of the eight patterns that can occur when a fair coin is tossed (Table 7.6).

Now let us play the coin game again where A is the first player and B is the second player. We show the probabilities that B beats A in Table 7.7.

Note that letting ">" means "beats" (probably) *TTH > HTT > HHT > THH > TTH*!

Table 7.7

A's Choice	B's Choice	P (B beats A)
HHH	*THH*	7/8
HHT	*THH*	3/4
HTH	*HHT*	2/3
HTT	*HHT*	2/3
THH	*TTH*	2/3
THT	*TTH*	2/3
TTH	*HTT*	3/4
TTT	*HTT*	7/8

Nor is it true that a pattern with a shorter average waiting time will necessarily beat a pattern with a longer waiting time. It can be shown that the average waiting time for *THTH* is 20 tosses and the average waiting time for *HTHH* is 18 tosses. Nonetheless, the probability that *THTH* occurs before *HTHH* is 9/14.

Probability contains many apparent contradictions.

WHO PAYS FOR LUNCH?

Three friends, whom we will call A, B, and C, go to lunch regularly. The payer at each lunch is selected randomly until someone pays for lunch for the second time. On average, how many times will the group go to lunch?

Let X denote the number of dinners the group enjoys. Clearly, $X = 2, 3$, or 4. We calculate the probabilities of each of these values.

If $X = 2$, then we have a choice of any of the three to pay for the first lunch. Then the same person must pay for the second lunch as well. The probability of this is $1/3$.

If $X = 3$, then we have a choice of any of the three to pay for the first lunch. Then we must choose one of the two who did not pay for the first lunch and finally, we must choose one of the two previous payers to pay for the third lunch. There are then $3 \cdot 2 \cdot 2 = 12$ ways in which this can be done and since each way has probability $1/27$, the probability that $X = 3$ is $12/27 = 4/9$.

Finally, if $X = 4$ then any of the three can pay for the first lunch; either of the other two must pay for the second lunch and the one remaining must pay for the third lunch. The fourth lunch can be paid by any of the three so this gives $3 \cdot 2 \cdot 1 \cdot 3 = 18$ ways in which this can be done. Since each has probability $(1/3)^4$, the probability that $X = 4$ is $18/81 = 2/9$.

These probabilities add up to 1 as they should.

The expected number of lunches is then $E(X) = 2 \cdot 3/9 + 3 \cdot 4/9 + 4 \cdot 2/9 = 26/9$. This is a bit under three, so they might as well go to lunch three times and forget the random choices except that sometimes someone never pays.

Now what happens as the size of the group increases? Does the randomness affect the number of lunches taken?

Suppose that there are four people in the group, A, B, C, and D. Then $X = 2, 3, 4$, or 5. We calculate the probabilities in much the same way as we did when there were three for lunch.

$$P(X = 2) = \frac{4 \cdot 1}{4 \cdot 4} = \frac{1}{4}$$

$$P(X = 3) = \frac{4 \cdot 3 \cdot 2}{4 \cdot 4 \cdot 4} = \frac{3}{8}$$

$$P(X = 4) = \frac{4 \cdot 3 \cdot 2 \cdot 3}{4 \cdot 4 \cdot 4 \cdot 4} = \frac{9}{32}$$

And finally,

$$P(X = 5) = \frac{4 \cdot 3 \cdot 2 \cdot 1 \cdot 4}{4 \cdot 4 \cdot 4 \cdot 4 \cdot 4} = \frac{3}{32}$$

and these sum to 1 as they should.

Then $E(X) = 2 \cdot 1/4 + 3 \cdot 3/8 + 4 \cdot 9/32 + 5 \cdot 3/32 = 103/32 = 3.218\,75$. This is now a bit under 4, so the randomness is having some effect.

To establish a general formula for $P(X = x)$ for n lunchers, note that the first payer can be any of the n people, the next must be one of the $n - 1$ people, the third one of the $n - 2$ people, and so on. The next to the last payer is one of the $n - (x - 2)$ people and the last payer must be one of the $x - 1$ people who have paid once.

This means that

$$P(X = x) = \frac{n}{n} \cdot \frac{n - 1}{n} \cdot \frac{n - 2}{n} \cdot \frac{n - 3}{n} \cdot \ldots \cdot \frac{n - (x - 2)}{n} \cdot \frac{x - 1}{n}$$

This can be rewritten as

$$P(X = x) = \frac{1}{n^x} \binom{n}{x - 1} (x - 1)(x - 1)!, \; x = 2, 3, \ldots, n + 1$$

If there are 10 diners, the probabilities of having x lunches are shown in Table 7.8.

Table 7.8

x	$P(X = x)$
2	0.1
3	0.18
4	0.216
5	0.2016
6	0.1512
7	0.09072
8	0.042336
9	0.0145152
10	0.00326592
11	0.00036288

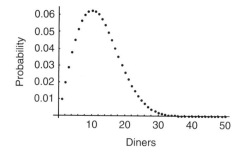

Figure 7.4

With a computer algebra system, it is possible to compute the probabilities for various values of X with some accuracy. In Figure 7.4 we show a graph of the probabilities for 100 diners.

EXPECTED NUMBER OF LUNCHES

It is also interesting to compute the expectations for increasing numbers of diners and to study the effect the randomness has. The computer algebra system *Mathematica* was used to produce Table 7.9 where n is the number of lunchers and the expected number of lunch parties is computed. The number of diners now increases beyond any sensible limit, providing some mathematical challenges.

Table 7.9

n	Expectation
2	2.500
3	2.889
4	3.219
5	3.510
6	3.775
10	4.660
15	5.546
20	6.294
30	7.550
50	9.543
100	13.210
150	16.025
200	18.398
300	22.381
400	25.738
450	27.258

We find

$$E[X] = \sum_{x=2}^{n+1} \frac{1}{n^x} \binom{n}{x-1} (x-1)x!$$

It is interesting to note that adding one person to the dinner party has less and less an effect as n increases. A graph of the expected number of lunches as a function of the number of diners is shown in Figure 7.5.

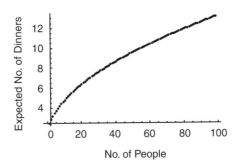

Figure 7.5

A least squares straight line regression gives *Expected No. of Lunches* $= 8.86266 + 0.04577 \cdot$ *No. of People*. A statistical test indicates that the fit is almost perfect for the calculated points.

NEGATIVE HYPERGEOMETRIC DISTRIBUTION

A manufacturer has a lot of 400 items, 50 of which are special. The items are inspected one at a time until 10 of the special items have been found. If the inspected items are not replaced in the lot, the random variable representing the number of special items found leads to the *negative hypergeometric distribution*.

The problem here is our final example of a *waiting time* problem. Had the inspected items been replaced, a questionable quality control procedure to say the least, we would encounter the *negative binomial distribution,* which we have seen when waiting for the rth success in a binomial process. Recall that if Y is the waiting time for the rth success then

$$P(Y = y) = \binom{y-1}{r-1} p^{r-1}(1-p)^{y-r} \cdot p, \ y = r, r+1, \cdots$$

We showed previously that the expected value of Y is $E(Y) = r/p$, a fact we will return to later.

Now we define the negative hypergeometric random variable. To be specific, suppose the lot of N items contains k special items. We want to sample, without replacement, until we find c of the special items. Then the sampling process stops. Again Y is the random variable denoting the number of trials necessary. Since the first

$y - 1$ trials comprise a hypergeometric process and the last trial must find a special item,

$$P(Y = y) = \frac{\binom{k}{c-1} \cdot \binom{N-k}{y-c}}{\binom{N}{y-1}} \cdot \frac{k - (c-1)}{N - (y-1)}, \quad y = c, c+1, \ldots, N - (k-c)$$

Note that the process can end in as few as c trials. The maximum number of trials must occur when the first $N - k$ trials contain all the nonspecial items followed by all c special items.

Some elementary simplifications show that

$$P(Y = y) = \frac{\binom{y-1}{c-1} \cdot \binom{N-y}{k-c}}{\binom{N}{k}}, \quad y = c, c+1, \ldots, N - (k-c)$$

In the special case we have been considering, $N = 400$, $k = 50$, and $c = 10$.

With a computer algebra system such as *Mathematica*®, we can easily calculate all the values in the probability distribution function and draw a graph that is shown in Figure 7.6.

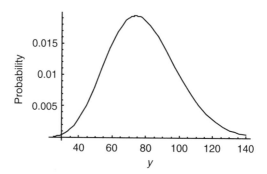

Figure 7.6

Some of the individual probabilities are shown in Table 7.10.

The mean value of Y is $E(Y) = 4010/51 = 78.6275$ and the variance is 425.454. Note that had we been using a negative binomial model the mean would be

$$\frac{c}{p} = \frac{c}{k/N} = \frac{10}{50/400} = 80$$

The fact that this negative binomial mean is always greater than that for the negative hypergeometric has some implications. This fact will be shown below, but first we establish formulas for the mean and variance of the negative hypergeometric random variable.

Table 7.10

y	Probability
40	0.00286975
45	0.00530886
50	0.00843214
55	0.0118532
60	0.0150666
65	0.0175912
70	0.0190899
75	0.0194303
80	0.0186812
85	0.0170618
90	0.0148697
95	0.0124115
100	0.00995123
105	0.00768271
110	0.00572262
115	0.00411924
120	0.0028691

MEAN AND VARIANCE OF THE NEGATIVE HYPERGEOMETRIC

We use a recursion. If we calculate $P(Y = y)/P(Y = y - 1)$, we find after simplification that

$$\sum_{y=c}^{N-k+c} (y - c)(N - y + 1)P(Y = y)$$

$$= \sum_{y=c}^{N-k+c} (y - 1)(N - y + 1 - k + c)P(Y = y - 1)$$

where we have also indicated a sum over all the possible values for Y.
This becomes

$$\sum_{y=c}^{N-k+c} [(N + 1 + c)y - y^2 - c(N + 1)]P(Y = y)$$

$$= \sum_{y=c}^{N-k+c} [(N - k + c)(y - 1) - (y - 1)^2]P(Y = y - 1)$$

and this can be further expanded and simplified to

$$(N + 1 + c)E(Y) - E(Y^2) - c(N + 1) = (N - k + c)[E(Y)$$
$$-(N - k + c)P(Y = N - k + c) - [E(Y^2) - (N - k + c)^2 P(Y = N - k + c)]$$

When this is simplified we find

$$E(Y) = \frac{c(N + 1)}{k + 1}$$

In our special case this gives $E(Y) = 10(400+1)/(50+1) = 4010/51 = 78.6275$.

Before proceeding to the variance, we establish the fact that the mean of the negative binomial always exceeds that of the negative hypergeometric.

Since $N > k$, $N + Nk > k + Nk$ so that $cN/k > c(N + 1)/k + 1$ establishing the result.

Although it is true that the negative binomial distribution is the limiting distribution for the negative hypergeometric distribution, this fact must be used sparingly if approximations are to be drawn. We will return to this point later.

Now we establish a formula for the variance of the negative hypergeometric random variable. We start with the previous recursion and multiply by y:

$$\sum_{y=c}^{N-k+c} y(y - c)(N - y + 1)P(Y = y)$$

$$= \sum_{y=c}^{N-k+c} y(y - 1)(N - y + 1 - k + c)P(Y = y - 1)$$

The left-hand side reduces quite easily to

$$(N + c + 1)E(Y^2) - E(Y^3) - c(N + 1)E(Y)$$

The right-hand side must be first expanded in terms of $y - 1$ and then it can be simplified to

$$(N + c - k)E(Y) + (N - 1 + c - k)E(Y^2) - E(Y^3)$$

Then, putting the sides together, we find

$$E(Y^2) = \frac{(cN + 2c + N - k)c(N + 1)}{(k + 2)(k + 1)}$$

from which it follows that

$$\text{Var}(Y) = \frac{c(N + 1)}{(k + 2)(k + 1)^2}[(N - k)(k - c + 1)]$$

In our special case, we find $\text{Var}(Y) = 425.454$ as before.

NEGATIVE BINOMIAL APPROXIMATION

It might be thought that the negative binomial distribution is a good approximation to the negative hypergeometric distribution. This is true as the values in Table 7.11 indicate.

Table 7.11

Y	Neghyper	Negbin	Neghyper − Negbin
90	0.014870	0.0135815	0.0012883
95	0.012412	0.011657	0.0007543
100	0.009951	0.009730	0.0002209
105	0.007683	0.007921	−0.000239
110	0.005723	0.006305	−0.000582
115	0.004119	0.004916	−0.000797
120	0.002869	0.003763	−0.000894
125	0.001936	0.002831	−0.000896
130	0.001266	0.002097	−0.000831
135	0.000803	0.001531	−0.000728
140	0.000494	0.001103	−0.000609
145	0.000295	0.000785	−0.000490
150	0.000171	0.000552	−0.000381
155	0.000096	0.000384	−0.000288
160	0.000053	0.000265	−0.000212

Figure 7.7 also shows that the values for the two probability distribution functions are very close.

It was noted above that the mean of the negative binomial distribution always exceeds that of the negative hypergeometric distribution. This difference is

$$\frac{c(N - k)}{k(k + 1)}$$

and this can lead to quite different results if probabilities are to be estimated.

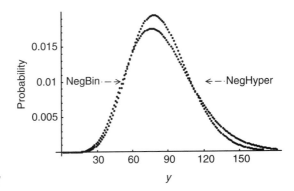

Figure 7.7

Figure 7.7 shows a comparison between the negative hypergeometric with $N = 400$, $k = 50$, and $c = 10$ and the negative binomial. The means differ by 1.3725 units. Although the graphs appear to be different, the actual differences are quite negligible as the values in Table 7.11, calculated from the right-hand tails, show.

The problem of course is the fact that the negative binomial assumes a constant probability of selecting a special item, while in fact this probability constantly changes with each item selected.

THE MEANING OF THE MEAN

We have seen that the expected value of Y is given by

$$E(Y) = \frac{c(N + 1)}{k + 1}$$

This has some interesting implications, one of which we now show.

First Occurrences

If we let $c = 1$ in $E(Y)$, we look at the expected number of drawings until the first special item is found. Table 7.12 shows some expected waiting times for a lot with $N = 1000$ and various values of k.

The graph in Figure 7.8 shows these results.

Waiting Time for c Special Items to Occur

Now consider a larger case. In $E(Y)$ let $N = 1000$ and $k = 100$. Table 7.13 shows the expected waiting time for c special items to occur.

Table 7.12

k	$E(Y)$
10	91.0000
20	47.6667
30	32.2903
40	24.4146
50	19.6275
60	16.4098
70	14.0986
80	12.3580
90	11.0000
100	9.91089

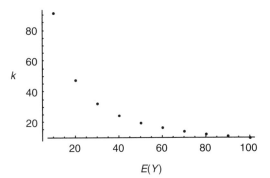

Figure 7.8 $E(Y)$

Note that, not surprisingly, if c is some percentage of k, then $E(Y)$ is approximately the same percentage of N, a result easily seen from the formula for $E(Y)$

The graph in Figure 7.9 shows this result as well.

Estimating k

Above we have assumed that k is known, a dubious assumption at best. On the basis of a sample, how can we estimate k? To be specific, suppose that a sample of 100 from a population of 400 shows 5 special items. What is the maximum likelihood estimate of k?

This is not a negative hypergeometric situation but a hypergeometric situation.

Consider $\text{prob}[k] = \dfrac{\binom{k}{5} \cdot \binom{400 - k}{95}}{\binom{400}{100}}$. The ratio $\text{prob}[k = 1]/\text{prob}[k]$ can be simplified to $((k + 1)(305 - k))/((k - 4)(400 - k))$. Seeing where this is > 1 gives $k < 19.05$.

Table 7.13

c	$E(Y)$
10	99.1089
20	198.218
30	297.327
40	396.436
50	495.545
60	594.653
70	693.762
80	792.871
90	891.980
100	991.089

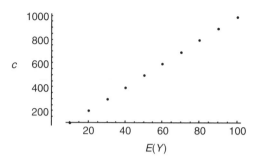

Figure 7.9

The sample then has 20% special items and that is somewhat greater than the maximum likelihood estimator for the percentage of special items in the population.

Now suppose the population is of size N, the sample is of size s, and the sample contains s_p special items. It is easy to show that $\hat{k}$ the maximum likelihood estimator for k, the unknown number if special items in the population, is

$$\hat{k} = \frac{s_p(N+1) - s}{s}$$

and further

$$\frac{\hat{k}}{N} = \frac{s_p}{s}\left(1 + \frac{1}{N}\right) - \frac{1}{N}$$

showing that the proportion of the population estimated is very close to the percentage of the special items in the sample, especially as the sample size increases, not really much of a surprise.

Here we have used the hypergeometric distribution to estimate k. We find exactly the same estimate if we use the negative hypergeometric waiting time distribution.

CONCLUSIONS

This chapter has contained a wide variety of waiting time problems. These are often not considered in introductory courses in probability and statistics and yet they offer interesting situations, both in their mathematical analysis and in their practical applications.

EXPLORATIONS

1. Show that if X_1 is the waiting time for the first binomial success and X_2 is the waiting time for the second binomial success, then $X_1 + X_2$ has a negative binomial distribution with $r = 2$.

2. Show that the expected waiting time for the pattern HT with a fair coin is four tosses.

3. **(a)** Suppose there are 4 people going for lunch as in the text. How many lunches could a person who never pays for lunch expect?

   (b) Repeat part (a) for a person who pays for lunch exactly once.

4. A lot of 50 items contains 5 defective items. Find the waiting times for the second defective item to occure if sampled items are not replaced before the next item is selected.

5. Two fair coins are tossed and any that comes up heads is put aside. This is repeated until all the coins have come up heads. Show that the expected number of (group) tosses is 8/3.

6. Professor Banach has two jars of candy on his desk and when a student visits, he or she is asked to select a jar and have a piece of candy. At some point, one of the jars is found to be empty. On average, how many pieces of candy are in the other jar?

7. **(a)** A coin, loaded to come up heads with probability 0.6, is thrown until a head appears. What is the probability an odd number of tosses is necessary?

   (b) If the coin is fair, explain why the probabilities of odd or even numbers of tosses are not equal.

8. Suppose X is a negative binomial random variable with p the probability of success at any trial. Suppose the rth success occurs at trial t. Find the value of p that makes this event most likely to occur.

Chapter 8

Continuous Probability Distributions: Sums, the Normal Distribution, and the Central Limit Theorem; Bivariate Random Variables

CHAPTER OBJECTIVES:

- to study random variables taking on values in a continuous interval or intervals
- to see how events with probability zero can and do occur
- to discover the surprising behavior of sums and means
- to use the normal distribution in a variety of settings
- to explain why the normal curve is called "normal"
- to discuss the central limit theorem
- to introduce bivariate random variables.

So far we have considered discrete random variables, that is, random variables defined on a discrete or countably infinite set. We now turn our attention to continuous random variables, that is, random variables defined on an infinite, uncountable, set such as an interval or intervals.

The simplest of these random variables is the *uniform* random variable.

A Probability and Statistics Companion, John J. Kinney
Copyright © 2009 by John Wiley & Sons, Inc.

UNIFORM RANDOM VARIABLE

Suppose we have a spinner on a wheel labeled with all the numbers between 0 and 1 on its circular border. The arrow is spun and the number the arrow stops at is the random variable, X. The random variable X can now take on *any* value in the interval from 0 to 1. What shall we take as the probability distribution function, $f(x) = P(X = x)$? We suppose the wheel is fair, that is, it is as likely to stop at any particular number as at any other. The wheel is shown in Figure 8.1.

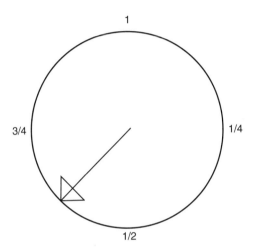

Figure 8.1

Clearly, we can not make $P(X = x)$ very large, since we have an infinity of values to use. Suppose we let

$$P(X = x) = 0.00000000000000000001 = 10^{-20}$$

The problem with this is that it can be shown that the wheel contains more than 10^{20} points, so we have used up more than the total probability of 1!

We are forced to conclude that $P(X = x) = 0$.

Now suppose that the wheel is loaded so that $P(X \geq 1/2) = 3P(X \geq 1/2)$, making it three times as likely that the arrow ends up in the left-hand half of the wheel as in the right-hand half of the wheel. What is $P(X = x)$ now?

Again, we conclude that $P(X = x) = 0$. It is curious that we cannot distinguish a loaded wheel from a fair one!

The difficulty here lies not in the answer to our question, but in the question itself. If we consider *any* random variable defined on a continuous interval, then $P(X = x)$ will always be 0. So we ask a different question, what is $P(a \leq x \leq b)$? That is, what is the probability that the random variable is contained in an interval? To make this meaningful, we define the probability *density* function, $f(x)$, which has the following properties:

1. $f(x) \geq 0$.

2. The total area under $f(x)$ is 1.

3. The area under $f(x)$ between $x = a$ and $x = b$ is $P(a \leq x \leq b)$.

So *areas* under $f(x)$ are probabilities. This means that $f(x)$ must always be positive, since probabilities cannot be negative. Also, $f(x)$ must enclose a total area of 1, the total probability for the sample space.

What is $f(x)$ for the fair wheel? Since, for the fair wheel, $P(a \leq x \leq b) = b - a$, it follows that

$$f(x) = 1, \quad 0 \leq x \leq 1.$$

Notice that $f(x) \geq 0$ and that the total area under $f(x)$ is 1. Here we call $f(x)$ a *uniform* probability density function. Its graph is shown in Figure 8.2.

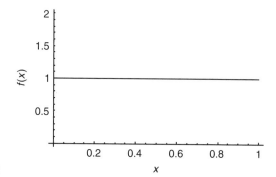

Figure 8.2

So, for example, to find $P(1/3 \leq X \leq 3/4)$, we find the area under the curve between 1/3 and 3/4. This is $(3/4 - 1/3) \cdot 1 = 5/12$. Since $f(x)$ is quite easy, areas under the curve, probabilities, are also easy to find.

For the loaded wheel, where $P(0 \leq X \leq 1/2) = 1/4$ and so $P(1/2 \leq X \leq 1) = 3/4$, consider (among many other choices) the triangular distribution

$$f(x) = 2x, \quad 0 \leq x \leq 1$$

The distribution is shown in Figure 8.3.

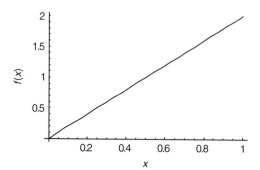

Figure 8.3

Areas can be found using triangles and it is easy to see, since $f(1/2) = 1$, that $P(0 \leq X \leq 1/2) = 1/4$.

SUMS

Now let us return to the fair wheel (where $f(x) = 1, 0 \leq x \leq 1$) and consider spinning the wheel twice and *adding* the numbers the pointer stops at. What is the probability density function for the sum? One might think that since we are adding two uniform random variables the sum will also be uniform, but that is not the case, as we saw in the discrete case in Chapter 5.

The sums obtained will then be between 0 and 2. Consider the probability of getting a small sum. For that to occur, *both* numbers must be small. Similarly, to get a sum near 2, a large sum, both spins must be near 1. So either of these possibilities is unlikely.

If X_1 and X_2 represent the outcomes on the individual spins, then the expected value of each is $E(X_1) = 1/2$ and $E(X_2) = 1/2$. While we cannot prove this in the continuous case, here is a fact and an example that may prove convincing.

A Fact About Means

If a probability distribution or a probability density function has a point of symmetry, then that point is the expected value of the random variable.

As an example, consider the discrete random variable X that assumes the values 1, 2, 3, 4, 5 with probabilities $f(1)$, $f(2)$, $f(3)$, $f(4)$, and $f(5)$, where $X = 3$ is a point of symmetry and $f(1) = f(5)$ and $f(2) = f(4)$. Now

$$E(X) = 1 \cdot f(1) + 2 \cdot f(2) + 3 \cdot f(3) + 4 \cdot f(4) + 5 \cdot f(5) = 6 \cdot f(1) + 6 \cdot f(2) + 3 \cdot f(3)$$

But

$$f(1) + f(2) + f(3) + f(4) + f(5) = 2 \cdot f(1) + 2 \cdot f(2) + f(3) = 1$$

so

$$6 \cdot f(1) + 6 \cdot f(2) + 3 \cdot f(3) = 3$$

the point of symmetry.

This is far from a general explanation or proof, but this approach can easily be generalized to a more general discrete probability distribution. We will continue to use this as a fact. It is also true for a continuous probability distribution; we cannot supply a proof here.

It is always true that $E(X_1 + X_2) = E(X_1) + E(X_2)$ (a fact we will prove later) and since $1/2$ is a point of symmetry, it follows that

$$E(X_1 + X_2) = 1/2 + 1/2 = 1$$

It is more likely to find a sum near 1 than to find a sum near either of the extreme values, namely, 0 or 2, as we shall see.

It can be shown that the probability distribution of $X = X_1 + X_2$ is

$$f_2(x) = \begin{cases} x & \text{for } 0 \le x \le 1 \\ 2 - x & \text{for } 1 \le x \le 2 \end{cases}$$

We should check that $f_2(x)$ is a probability density function. The function is always positive. The easiest way to find its total area is to use the area of two triangles. Each of these have base of length 1 and height 1 as well, so the total area is $2 \cdot 1/2 \cdot 1 \cdot 1 = 1$; so $f_2(x)$ is a probability density function.

A graph of $f_2(x)$ is shown in Figure 8.4.

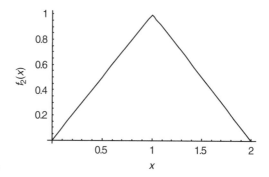

Figure 8.4

Since areas represent probabilities, using the area of a triangle, we find that $P(1/2 \le X \le 3/2) = 3/4$, so the sums do in fact cluster around their expected value.

If we increase the number of spins to 3 and record the sum, it can be shown that the probability density function for $X = X_1 + X_2 + X_3$ is

$$f_3(x) = \begin{cases} \dfrac{1}{2}x^2 & \text{for } 0 \le x \le 1 \\ \dfrac{3}{4} - \left(x - \dfrac{3}{2}\right)^2 & \text{for } 1 \le x \le 2 \\ \dfrac{1}{2}(x - 3)^2 & \text{for } 2 \le x \le 3 \end{cases}$$

This is also a probability density function, although finding the total area, or finding probabilities by finding areas, is usually done using calculus. We can find here, for example, that $P(1/2 \le x \le 5/2) = 23/24$. Probabilities, or areas, are now difficult for us to find and we cannot do this by simple geometry. However, we will soon find a remarkable approximation to this probability density function that will enable us to determine areas and hence probabilities .

A graph of $f_3(x)$ is shown in Figure 8.5.

The graph in Figure 8.5 resembles a "bell-shaped" or *normal* curve that occurs frequently in probability and statistics. We have encountered this curve before.

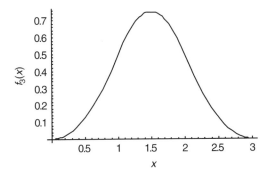

Figure 8.5

If we were to continue adding spins, we would find that the graphs come closer and closer to a *normal* curve. This is an illustration of the *central limit theorem*. We will discuss the normal curve first and then state the central limit theorem.

NORMAL PROBABILITY DISTRIBUTION

The *normal probability distribution* is a bell-shaped curve that is completely specified by its mean μ and its standard deviation σ. Its probability density function is

$$f(x) = \frac{1}{\sigma\sqrt{2\pi}}e^{-\frac{1}{2}(x-\mu)^2}, \quad -\infty \le \mu \le \infty, \quad \sigma \ge 0, \quad -\infty \le x \le \infty$$

Note that for any normal curve, the curve is symmetric about its mean value.

A typical normal distribution is shown in Figure 8.6, which is a normal curve with mean 3 and standard deviation $\sqrt{2}$.

We abbreviate any normal curve by specifying its mean and standard deviation and we write, for a general normal random variable X, $X \sim N(\mu, \sigma)$. This is read, "X is distributed normally with mean μ and standard deviation σ". For the normal curve

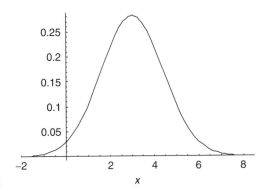

Figure 8.6

in Figure 8.6, we write $X \sim N(3, \sqrt{2})$. It can be shown, but with some difficulty, that the total area of any normal curve is 1. Since the function is always positive, areas under the curve represent probabilities. These cannot be calculated easily either.

Statistical calculators, however, can calculate these areas. Here are some examples from this normal curve.

(a) $P(2 \leq X \leq 6) = 0.743303$

(b) $P(4 \leq X \leq 8) = 0.239547$

(c) $P(X > 4 | X > 2) = \dfrac{P(X > 4 \text{ and } X > 2)}{P(X > 2)} = \dfrac{P(X > 4)}{P(X > 2)} = \dfrac{0.23975}{0.76025}$
$= 0.315357$

Textbooks commonly include a table of the *standard* normal curve, that is, $N(0, 1)$. Since computers and calculators compute areas under the standard normal curve, we do not need to include such a table in this book. The use of this standard normal curve to calculate areas under any normal curve is based on this fact.

Fact.   If $X \sim N(\mu, \sigma)$ and if $Z = (X - \mu)/\sigma$, then

$Z \sim N(0, 1)$.

This means that areas under any normal curve can be calculated using a single, standard, normal curve. For example, using our example (a) above,

$$P(2 \leq X \leq 6) = P\left(\frac{2 - 3}{\sqrt{2}} = \frac{X - 3}{\sqrt{2}} = \frac{6 - 3}{\sqrt{2}}\right)$$
$$= P(-0.707\,11 \leq Z \leq 2.\,121\,3) = 0.743303$$

as before. The fact that any normal curve can be transformed into a standard normal curve is a remarkable fact (and a fact not true for the other probability density functions we will encounter). There are, however, many other uses of the standard normal curve and we will meet these when we study other statistical topics.

Facts About Normal Curves

To show that the standard deviation does in fact measure the dispersion of the distribution, we calculate several probabilities for a standard $N(0, 1)$ distribution.

(a) $P(-1 \leq Z \leq 1) = 0.6827$

(b) $P(-2 \leq Z \leq 2) = 0.9545$

(c) $P(-3 \leq Z \leq 3) = 0.9973$

So the more standard deviations we use, the more of the distribution we enclose.

EXAMPLE 8.1   *IQ Scores*

A standard test for the intelligence quotient (IQ) produces scores that are approximately normally distributed with mean 100 and standard deviation 10. So, using the facts above and assuming $Z = (IQ - 100)/10$, we see that $P(90 \leq IQ \leq 110) = 0.6827$, $P(80 \leq IQ \leq 120) = 0.9545$; and $P(70 \leq IQ \leq 130) = 0.9973$.

Also we can calculate the percentage of the population whose IQ values are greater than 140, as $P(IQ \geq 140) = 1 - P(Z \leq 4) = 1 - 0.9968 = 0.0032$, a rarity. However, in a population of about $300,000,000$ in the United States, this gives about $1,000,000$ people with this IQ or greater.   ■

There are many other applications for the normal distribution. One of the reasons for this, but not the only one, is that sums of different random variables or means of several random variables become normal. We had an indication of this when we added uniformly distributed random variables. This is a consequence of the *central limit theorem*, which we will discuss subsequently.

But before we do that, we turn our attention to *bivariate* random variables.

BIVARIATE RANDOM VARIABLES

We want to look at sums when we add the observations from the spinning wheel. We have been concerned with a single random variable, but first note that the observations we want to add are those from different observations and so are *different* random variables. So we turn our attention to different random variables and first consider two different random variables.

Suppose we have a sample space and we have defined two random variables, which we will call X and Y, on the sample points. Now we must determine the probabilities that the random variables assume values *together*. We need not show the sample space, but the probabilities with which the random variables take on their respective values together are shown in Table 8.1. This is called the *joint probability distribution function*.

The table is to be read this way: X can take on the values 1, 2, and 3 while the random variable Y can take on the values 1 and 2. The entries in the body of the table

Table 8.1

| | | X | |
Y	1	2	3
1	1/12	1/12	1/3
2	1/3	1/12	1/12

are the probabilities that X and Y assume their values simultaneously. For example,

$$P(X = 1 \text{ and } Y = 2) = 1/3 \quad \text{and} \quad P(X = 3 \text{ and } Y = 2) = 1/12$$

Note that the probabilities in the table add up to 1 as they should.

Now consider the random variable $X + Y$. This random variable can take on the values 2, 3, 4, or 5.

There is only one way for $X + Y$ to be 2, namely, each of the variables must be 1. So the probability that $X + Y = 2$ is $1/12$ and we write $P(X + Y = 2) = 1/12$.

There are two mutually exclusive ways for $X + Y$ to be 3, namely, $X = 1$ and $Y = 2$ or $X = 2$ and $Y = 1$. So $P(X + Y = 3) = P(X = 1 \text{ and } Y = 2) + P(X = 2 \text{ and } Y = 1) = 1/3 + 1/12 = 5/12$.

It is easy to check that there are two mutually exclusive ways for $X + Y$ to be 4 and this has probability $5/12$. Finally, the probability that $X + Y = 5$ is $1/12$. These probabilities add up to 1 as they should.

This means that the random variable $X + Y$ has the following probability distribution function:

$$f(x + y) = \begin{cases} 1/12 & \text{if } x + y = 2 \\ 5/12 & \text{if } x + y = 3 \\ 5/12 & \text{if } x + y = 4 \\ 1/12 & \text{if } x + y = 5 \end{cases}$$

where x and y denote values of the random variables X and Y.

This random variable then has a mean value. We find that

$$E(X + Y) = 2 \cdot \frac{1}{12} + 3 \cdot \frac{5}{12} + 4 \cdot \frac{5}{12} + 5 \cdot \frac{1}{12} = \frac{7}{2}$$

How does this value relate to the expected values of the variables X and Y taken separately?

First, we must find the probability distribution functions of the variables alone. What, for example, is the probability that $X = 1$? We know that $P(X = 1 \text{ and } Y = 1) = 1/12$ and $P(X = 1 \text{ and } Y = 2) = 1/3$. These events are mutually exclusive and are the only events for which $X = 1$. So

$$P(X = 1) = 1/12 + 1/3 = 5/12$$

Notice that this is the sum of the probabilities in the column in Table 8.1 for which $X = 1$.

In a similar way,

$$P(X = 2) = P(X = 2 \text{ and } Y = 1) + P(X = 2 \text{ and } Y = 2) = 1/12 + 1/12 = 1/6,$$

which is the sum of the probabilities in the column of Table 8.1 for which $X = 2$

Finally, summing the probabilities in Table 8.1 for which $X = 3$ is $1/3 + 1/12 = 5/12$.

So the probability distribution for the random variable X alone can be found by adding up the entries in the columns; the probability distribution for the random variable Y alone can be found by adding up the probabilities in the rows of Table 8.1.

Specifically,

$$P(Y = 1) = P(X = 1 \text{ and } Y = 1) + P(X = 2 \text{ and } Y = 1) + P(X = 3 \text{ and } Y = 1)$$
$$= 1/12 + 1/12 + 1/3 = 1/2$$

In a similar way, we find $P(Y = 2) = 1/2$.
We then found the following probability distributions for the individual variables:

$$f(x) = \begin{cases} 5/12 & \text{if } x = 1 \\ 1/6 & \text{if } x = 2 \\ 5/12 & \text{if } x = 3 \end{cases}$$

and

$$g(y) = \begin{cases} 1/2 & \text{if } y = 1 \\ 1/2 & \text{if } y = 2 \end{cases}$$

These distributions occur in the margins of the table and are called *marginal distributions.* We have expanded Table 8.1 to show these marginal distributions in Table 8.2.

Table 8.2

		X		
Y	1	2	3	$g(y)$
1	1/12	1/12	1/3	1/2
2	1/3	1/12	1/12	1/2
$f(x)$	5/12	1/6	5/12	1

Note that where the sums are over all the values of X and Y,

$$\sum_x P(X = x, Y = y) = P(Y = y) = g(y)$$

and

$$\sum_y P(X = x, Y = y) = P(X = x) = f(x)$$

These random variables also have expected values. We find

$$E(X) = 1 \cdot \frac{5}{12} + 2 \cdot \frac{1}{6} + 3 \cdot \frac{5}{12} = 2$$

$$E(Y) = 1 \cdot \frac{1}{2} + 2 \cdot \frac{1}{2} = \frac{3}{2}$$

Now we note that $E(X + Y) = \frac{7}{2} = 2 + \frac{3}{2} = E(X) + E(Y)$.

This is not a peculiarity of this special case, but is, in fact, true for any two variables X and Y. Here is a proof.

$$E(X + Y) = \sum_x \sum_y (x + y) P(X = x, \, Y = y)$$

$$= \sum_x \sum_y x P(X = x, \, Y = y) + \sum_x \sum_y y P(X = x, \, Y = y)$$

$$= \sum_x x \sum_y P(X = x, \, Y = y) + \sum_y y \sum_x P(X = x, \, Y = y)$$

$$= \sum_x x f(x) + \sum_y y \, g(y) = E(X) + E(Y)$$

This is easily extended to any number of random variables:

$$E(X + Y + Z + \cdots) = E(X) + E(Y) + E(Z) + \cdots$$

When more than one random variable is defined on the same sample space, they may be related in several ways: they may be totally *dependent* as, for example, if $X = Y$ or if $X = Y - 4$, they may be totally *independent* of each other, or they may be *partially dependent* on each other. In the latter case, the variables are called *correlated*. This will be dealt with when we consider the subject of regression later. It is important to emphasize that

$$E(X + Y + Z + \cdots) = E(X) + E(Y) + E(Z) + \cdots$$

no matter what the relationships are between the several variables, since no condition was used in the proof above.

Note in the example we have been considering that

$$P(X = 1 \text{ and } Y = 1) = 1/12 \neq P(X = 1) \cdot P(Y = 1) = 5/12 \cdot 1/2 = 5/24$$

so X and Y are *not* independent.

Now consider an example where X and Y are independent of each other. We show another joint probability distribution function in Table 8.3.

Table 8.3

y	X 1	2	3	$g(y)$
1	5/24	1/12	5/24	1/2
2	5/24	1/12	5/24	1/2
$f(x)$	5/12	1/6	5/12	1

Note that $P(X = x, Y = y) = P(X = x)P(Y = y)$ in each case, so the random variables are independent. We can calculate, for example,

$$P(X = 1 \text{ and } Y = 1) = 5/24 = P(X = 1) \cdot P(Y = 1) = 1/12 \cdot 1/2$$

The other entries in the table can be checked similarly. Here we have shown the marginal distributions of the random variables X and Y , $f(x)$ and $g(y)$, in the margins of the table.

Now consider the random variable $X \cdot Y$ and in particular its expected value. Using the fact that X and Y are independent, we sum the values of $X \cdot Y$ multiplied by their probabilities to find the expected value of the product of X and Y:

$$E(X \cdot Y) = 1 \cdot 1 \cdot \frac{1}{2} \cdot \frac{5}{12} + 1 \cdot 2 \cdot \frac{1}{6} \cdot \frac{1}{2} + 1 \cdot 3 \cdot \frac{5}{12} \cdot \frac{1}{2} + 2 \cdot 1 \cdot \frac{5}{12} \cdot \frac{1}{2}$$

$$+ 2 \cdot 2 \cdot \frac{1}{6} \cdot \frac{1}{2} + 2 \cdot 3 \cdot \frac{5}{12} \cdot \frac{1}{2}$$

$$= 3$$

but we also see that this can be written as

$$E(X \cdot Y) = \left(1 \cdot \frac{5}{12} + 2 \cdot \frac{1}{6} + 3 \cdot \frac{5}{12} \right) \cdot \left(1 \cdot \frac{1}{2} + 2 \cdot \frac{1}{2} \right)$$

$$= 2 \cdot \frac{3}{2} = 3 = E(X)E(Y)$$

and we see that the quantities in parentheses are $E(X)$ and $E(Y)$, respectively.

This is true of the general case, that is, if X and Y are independent, then

$$E(X \cdot Y) = E(X) \cdot E(Y)$$

A proof can be fashioned by generalizing the example above.

This can be extended to any number of random variables: if $X, Y, Z, \cdots$ are mutually independent in pairs, then

$$E(X \cdot Y \cdot Z \cdots) = E(X) \cdot E(Y) \cdot E(Z) \cdots$$

Although the examples given here involve discrete random variables, the results concerning the expected values are true for continuous random variables as well.

Variance

The variance of a random variable, X, is

$$\text{Var}(X) = E(X - \mu)^2$$

where $\mu = E(X)$.

This definition holds for both discrete and continuous random variables. Now

$$E(X - \mu)^2 = E(X^2 - 2\mu X + \mu^2)$$

$$= E(X^2) - 2\mu E(X) + E(\mu^2)$$

and since μ is a constant,

$$E(X - \mu)^2 = E(X^2) - 2\mu^2 + \mu^2$$

so

$$E(X - \mu)^2 = E(X^2) - \mu^2$$

The variance is a measure of the dispersion of a random variable as we have seen with the normal random variable, but, as we will show when we study the design of experiments, it can often be partitioned into parts that explain the source of the variation in experimental results.

We conclude our consideration of expected values with a result concerning the variance of a sum of independent random variables. By definition,

$$\begin{aligned}
\text{Var}(X + Y) &= E\left[(X + Y) - (\mu_x + \mu_y)\right]^2 \\
&= E\left[(X - \mu_x) + (Y - \mu_y)\right]^2 \\
&= E(X - \mu_x)^2 - 2E(X - \mu_x)(Y - \mu_y) + E(Y - \mu_y)^2
\end{aligned}$$

Consider the middle term. Since X and Y are independent,

$$\begin{aligned}
E(X - \mu_x)(Y - \mu_y) &= E(X \cdot Y - \mu_x \cdot Y - X \cdot \mu_y + \mu_x \cdot \mu_y) \\
&= E(X \cdot Y) - \mu_x \cdot E(Y) - E(X) \cdot \mu_y + E(\mu_x \cdot \mu_y) \\
&= E(X \cdot Y) - \mu_x \cdot \mu_y - \mu_x \cdot \mu_y + \mu_x \cdot \mu_y \\
&= E(X) \cdot E(Y) - \mu_x \cdot \mu_y - \mu_x \cdot \mu_y + \mu_x \cdot \mu_y \\
&= \mu_x \cdot \mu_y - \mu_x \cdot \mu_y = 0
\end{aligned}$$

So

$$\begin{aligned}
\text{Var}(X + Y) &= E(X - \mu_x)^2 + E(Y - \mu_y)^2 \\
&= \text{Var}(X) + \text{Var}(Y)
\end{aligned}$$

Note that this result highly depends on the independence of the random variables. As an example, consider the marginal distribution functions given in Table 8.3 and repeated here in Table 8.4.

Table 8.4

		X		
Y	1	2	3	$g(y)$
1	5/24	1/12	5/24	1/2
2	5/24	1/12	5/24	1/2
$f(x)$	5/12	1/6	5/12	1

We find that

$$E(X+Y) = 2 \cdot \frac{5}{24} + 3 \cdot \frac{7}{24} + 4 \cdot \frac{7}{24} + 5 \cdot \frac{5}{24} = \frac{84}{24} = \frac{7}{2}$$

So

$$Var(X+Y) = \left(2 - \frac{7}{2}\right)^2 \cdot \frac{5}{24} + \left(3 - \frac{7}{2}\right)^2 \cdot \frac{7}{24} + \left(4 - \frac{7}{2}\right)^2 \cdot \frac{7}{24}$$
$$+ \left(5 - \frac{7}{2}\right)^2 \cdot \frac{5}{24} = \frac{13}{12}$$

But

$$E(X) = 1 \cdot \frac{5}{12} + 2 \cdot \frac{1}{6} + 3 \cdot \frac{5}{12} = 2$$

and

$$E(Y) = 1 \cdot \frac{1}{2} + 2 \cdot \frac{1}{2} = \frac{3}{2}$$

so

$$Var(X) = (1-2)^2 \cdot \frac{5}{12} + (2-2)^2 \cdot \frac{1}{6} + (3-2)^2 \cdot \frac{5}{12} = \frac{5}{6}$$

and

$$Var(Y) = \left(1 - \frac{3}{2}\right)^2 \cdot \frac{1}{2} + \left(2 - \frac{3}{2}\right)^2 \cdot \frac{1}{2} = \frac{1}{4}$$

and so

$$Var(X) + Var(Y) = \frac{5}{6} + \frac{1}{4} = \frac{13}{12}$$

We could also calculate the variance of the sum by using the formula

$$Var(X+Y) = E(X+Y)^2 - [E(X+Y)]^2$$

Here

$$E(X+Y)^2 = 2^2 \cdot \frac{5}{24} + 3^2 \cdot \frac{7}{24} + 4^2 \cdot \frac{7}{24} + 5^2 \cdot \frac{5}{24} = \frac{40}{3}$$

We previously calculated $E(X+Y) = 7/2$, so we find

$$Var(X+Y) = \frac{40}{3} - \left(\frac{7}{2}\right)^2 = \frac{13}{12}$$

as before.

Now we can return to the spinning wheel.

CENTRAL LIMIT THEOREM: SUMS

We have seen previously that the bell-shaped curve arises when discrete random variables are added together. Now we look at continuous random variables. Suppose that we have n independent spins of the fair wheel, denoted by X_i, and we let the random variable X denote the sum so that

$$X = X_1 + X_2 + X_3 + \cdots + X_n$$

For the individual observations, we know that the expected value is $E(X_i) = 1/2$ and the variance can be shown to be $\text{Var}(X_i) = 1/12$. In addition, we know that

$$
\begin{aligned}
E(X) &= E(X_1 + X_2 + X_3 + \cdots + X_n) \\
&= E(X_1) + E(X_2) + E(X_3) + \cdots + E(X_n) \\
&= 1/2 + 1/2 + 1/2 + \cdots + 1/2 \\
&= n/2
\end{aligned}
$$

and since the spins are independent,

$$
\begin{aligned}
\text{Var}(X) &= \text{Var}(X_1 + X_2 + X_3 + \cdots + X_n) \\
&= \text{Var}(X_1) + \text{Var}(X_2) + \text{Var}(X_3) + \cdots + \text{Var}(X_n) \\
&= 1/12 + 1/12 + 1/12 + \cdots + 1/12 \\
&= n/12
\end{aligned}
$$

The *central limit theorem* states that in this case, X has, approximately, a normal probability distribution with mean $n/2$ and standard deviation $\sqrt{n/12}$. We abbreviate this by writing $X \backsim N(n/2, \sqrt{n/12})$. If $n = 3$, this becomes

$$X \backsim N(3/2, 1/2)$$

The value of n in the central limit theorem, as we have already stated, need not be very large. To show how close the approximation is, we show the graph $N(3/2, 1/2)$ in Figure 8.7 and then, in Figure 8.8, the graphs of $N(3/2, 1/2)$ and $f_3(x)$ superimposed.

We previously calculated, using $f_3(x)$, that $P(1/2 \le x \le 5/2) = 23/24 = 0.958\,33$. Using the normal curve with mean $3/2$ and standard deviation $1/2$, we find the

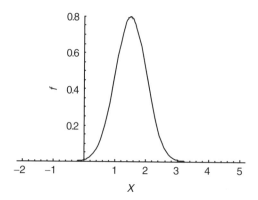

Figure 8.7

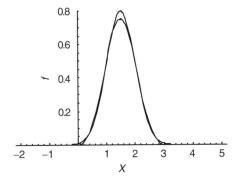

Figure 8.8

approximation to this probability to be 0.95450. As the number of spins increases, the approximation becomes better and better.

CENTRAL LIMIT THEOREM: MEANS

Understanding the central limit theorem is absolutely essential to understanding the material on statistical inference that follows in this book. In many instances, we know the mean from a random sample and we wish to make some inference or draw some conclusion, about the mean of the population from which the sample was selected. So we turn our attention to means.

First, suppose X is some random variable and k is a constant. Then, supposing that X is a discrete random variable,

$$E\left(\frac{X}{k}\right) = \sum_{S} \left(\frac{x}{k}\right) P(X = x)$$

$$= \left(\frac{1}{k}\right) \sum_{S} x \cdot P(X = x)$$

$$= \left(\frac{1}{k}\right) E(X)$$

Here the summation is over all the values in the sample space, S. Therefore, if the variable is divided by a constant, so is the expected value. Now, denoting $E(X)$ by μ,

$$\text{Var}\left(\frac{X}{k}\right) = \sum_{S} \left(\frac{x}{k} - \frac{\mu}{k}\right)^2 P(X = x)$$

$$= \left(\frac{1}{k^2}\right) \sum_{S} (x - \mu)^2 P(X = x)$$

$$= \left(\frac{1}{k^2}\right) \text{Var}(X)$$

The divisor, k, this time reduces the variance by its square.

CENTRAL LIMIT THEOREM

If $\overline{X}$ denotes the mean of a sample of size n from a probability density function with mean μ and standard deviation σ, then $\overline{X} \sim N(\mu, \sigma/\sqrt{n})$.

This theorem is the basis on which much of statistical inference, our ability to draw conclusions from samples and experimental data, is based. Statistical inference is the subject of the next two chapters.

EXPECTED VALUES AND BIVARIATE RANDOM VARIABLES

We now expand our knowledge of bivariate random variables. But before we can discuss the distribution of sample means, we pause to consider the calculation of means and variances of means.

Means and Variances of Means

We will now discuss the distribution of sample means. Suppose, as before, that we have a sum of independent random variables so that $X = X_1 + X_2 + X_3 + \cdots + X_n$. Suppose also that for each of these random variables, $E(X_i) = \mu$ and $\mathrm{Var}(X_i) = \sigma^2$.

The mean of these random variables is

$$\overline{X} = \frac{X_1 + X_2 + X_3 + \cdots + X_n}{n}$$

Using the facts we just established, we find that

$$
\begin{aligned}
E(\overline{X}) &= \frac{E(X_1 + X_2 + X_3 + \cdots + X_n)}{n} \\
&= \frac{E(X_1) + E(X_2) + E(X_3) + \cdots + E(X_n)}{n} \\
&= \frac{n\mu}{n} = \mu
\end{aligned}
$$

So the expected value of the mean of a number of random variables with the same mean is the mean of the individual random variables.

We also find that

$$
\begin{aligned}
\mathrm{Var}(\overline{X}) &= \frac{\mathrm{Var}(X_1 + X_2 + X_3 + \cdots + X_n)}{n^2} \\
&= \frac{\mathrm{Var}(X_1) + \mathrm{Var}(X_2) + \mathrm{Var}(X_3) + \cdots + \mathrm{Var}(X_n)}{n^2} \\
&= \frac{n\sigma^2}{n^2} = \frac{\sigma^2}{n}
\end{aligned}
$$

While the mean of the sample means is the mean of the distribution of the individual X_i's, the variance is reduced by a factor of n. This shows that the larger the sample size, the smaller the $\text{Var}(\overline{X})$. This has important implications for sampling. We show some examples.

EXAMPLE 8.2   *Means of Random Variables*

(a) The probability that an individual observation from the uniform random variable with $f(x) = 1$ for $0 \le x \le 1$ is between $1/3$ and $2/3$ is $(2/3 - 1/3) = 1/3$.

What is the probability that the mean of a sample of 12 observations from this distribution is between $1/3$ and $2/3$?

For the uniform random variable, $\mu = 1/2$ and $\sigma^2 = 1/12$; so, using the central limit theorem for means, we know that $E(\overline{X}) = 1/2$ and $\text{Var}(\overline{X}) = \text{Var}(x)/n = (1/12)/12 = 1/144$. This means that the standard deviation of $\overline{X}$ is $1/12$.

The central limit theorem for means then states that $\overline{X} \sim N(1/2, 1/12)$. Using a statistical calculator, we find

$$P(1/3 \le \overline{X} \le 2/3) = 0.9545$$

So, while an individual observation falls between $1/3$ and $2/3$ only about $1/3$ of the time, the sample mean is almost certain to do so.

(b) How large a sample must be selected from a population with mean 10 and standard deviation 2 so that the probability that the sample mean is within 1 unit of the population mean is 0.95?

Let the sample size be n. We know that $\overline{X} \sim N(10, 2/\sqrt{n})$. We want n so that

$$P(9 \le \overline{X} \le 11) = 0.95$$

We let $Z = (\overline{X} - 10)/(2/\sqrt{n})$. So we have

$$P\left(\frac{9 - 10}{2/\sqrt{n}} \le \frac{\overline{X} - 10}{2/\sqrt{n}} \le \frac{11 - 10}{2/\sqrt{n}}\right) = 0.95$$

or

$$P\left(\frac{-1}{2/\sqrt{n}} \le Z \le \frac{1}{2/\sqrt{n}}\right) = 0.95$$

But we know, for a standard normal variable, that $P(-1.96 \le Z \le 1.96) = 0.95$, so we conclude that

$$\frac{1}{2/\sqrt{n}} = 1.96$$

or

$$\sqrt{n}/2 = 1.96$$

so

$$\sqrt{n} = 2 \cdot 1.96 = 3.92$$

so

$$n = (3.92)^2 = 15.366$$

meaning that a sample of 16 is necessary. If we were to round down to 15, the probability would be less than 0.95.   ∎

A NOTE ON THE UNIFORM DISTRIBUTION

We have stated that for the uniform distribution, $E(X) = 1/2$ and $\text{Var}(X) = 1/12$, but we have not proved this. The reason is that these calculations for a continuous random variable require calculus, and that is not a prerequisite here. We give an example that may be convincing.

Suppose we have a discrete random variable, X, where $P(X = x) = 1/100$ for $x = 0.01, 0.02, \ldots, 1.00$. This is a discrete approximation to the continuous uniform distribution. We will need the following formulas:

$$1 + 2 + \cdots + n = \sum_{i=1}^{n} i = \frac{n(n+1)}{2}$$

and

$$1^2 + 2^2 + \cdots + n^2 = \sum_{i=1}^{n} i^2 = \frac{n(n+1)(2n+1)}{6}$$

We know that for a discrete random variable,

$$E(X) = \sum_{S} x \cdot P(X = x)$$

which becomes in this case

$$E(X) = \sum_{x=0.01}^{1.00} x \cdot (1/100) = (1/100) \cdot \sum_{x=0.01}^{1.00} x$$

Now assuming $x = i/100$ allows the variable i to assume integer values

$$E(X) = (1/100) \cdot \sum_{i=1}^{100} \left(\frac{i}{100} \right)$$

$$= (1/100)^2 \cdot \sum_{i=1}^{100} i$$

$$= (1/100)^2 \cdot \frac{(100)(101)}{2}$$

$$= 0.505$$

To calculate the variance, we first find $E(X^2)$. This is

$$E(X^2) = \sum_{x=0.01}^{1.00} x^2 \cdot P(X = x)$$

$$= \sum_{x=0.01}^{1.00} x^2 \cdot (1/100)$$

$$= (1/100) \cdot \sum_{x=0.01}^{1.00} x^2$$

and again assuming $x = i/100$,

$$E(X^2) = (1/100) \cdot \sum_{i=1}^{100} \left(\frac{i}{100}\right)^2$$

$$= (1/100)^3 \cdot \sum_{i=1}^{100} i^2$$

$$= (1/100)^3 \cdot \frac{(100)(101)(201)}{6}$$

$$= 0.33835$$

Now   it   follows   that
$\text{Var}(X) = E(X^2) - [E(X)]^2 = 0.33835 - (0.505)^2 = 0.083325 \simeq 1/12$.

If we were to increase the number of subdivisions in the interval from 0 to 1 to $10,000$ (with each point then having probability 0.0001) we find that

$$E(X) = 0.50005$$

and

$$\text{Var}(X) = 0.08333333325$$

This may offer some indication, for the continuous uniform random variable, that $E(X) = 1/2$ and $\text{Var}(X) = 1/12$.

We conclude this chapter with an approximation of a probability from a proba-bility density function.

EXAMPLE 8.3   *Areas Without Calculus*

Here is a method by which probabilities—areas under a continuous probability density function—can be approximated without using calculus.

Suppose we have *part* of a probability density function that is part of the parabola

$$f(x) = \frac{1}{2}x^2, \quad 0 \leq x \leq 1$$

and we wish to approximate $P(0 \leq x \leq 1)$. The exact value of this probability is $1/6$. The graph of this function is shown in Figure 8.9.

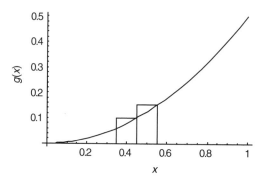

Figure 8.9

We will approximate the probability, or the area under the curve, by a series of rectangles, each of width 0.1, using the right-hand ends of the rectangles at the points $0.1, 0.2, \ldots, 1$. We use the height of the curve at the midpoints of these rectangles and use the total area of these rectangles as an approximation to the area, A, under the curve. Two of the approximating rectangles are shown in Figure 8.9. This gives us

$$A = (0.1) \cdot \frac{1}{2} \cdot (0.05)^2 + (0.1) \cdot \frac{1}{2} \cdot (0.15)^2$$

$$+ (0.1) \cdot \frac{1}{2} \cdot (0.25)^2 + \cdots + (0.1) \cdot \frac{1}{2} \cdot (0.95)^2 = 0.16625$$

which is not a bad approximation to the actual area, $0.1666\ldots$ ∎

This is a common technique in calculus, but the approximation here is surprisingly good. Increasing the number of rectangles only improves the approximation.

CONCLUSIONS

This chapter introduces continuous random variables, that is, random variables that can assume values in an interval or intervals.

We have found some surprising things when we added independent random variables, producing the normal probability distribution. We then stated the central limit theorem, a basic result when we study statistical inference in succeeding chapters.

We discussed some methods for approximating continuous distributions by discrete distributions, adding some credence to the statements we have made without proof for continuous distributions.

EXPLORATIONS

1. Mathematics scores on the Scholastic Aptitude Test (SAT) are normally distributed with mean 500 and standard deviation 100.
   (a) Find the probability that an individual's score exceeds 620.
   (b) Find the probability that an individual's score exceeds 620, given that the individual's score exceeds 500.
   (c) What score, or greater, can we expect to occur with probability 0.90?

2. A manufactured part is useful only if a measurement is between 0.25 and 0.38 in. The measurements follow a normal distribution with mean 0.30 and standard deviation 0.03 in.
   (a) What proportion of the parts meet specifications?
   (b) Suppose the mean measurement could be changed, but the standard deviation cannot be changed. To what value should the mean be changed to maximize the proportion of parts that meet specifications?

3. Upper and lower *warning limits* are often set for manufactured products. If $X \sim N(\mu, \sigma)$, these are commonly set at $\mu \pm 1.96\sigma$. If the mean of the process increases by one standard deviation, what effect does this have on the proportion of parts outside the warning limits?

4. Suppose

$$f(x) = \begin{cases} x, & 0 \le x \le 1 \\ 2 - x, & 1 \le x \le 2 \end{cases}$$

   (X is the sum of two uniformly distribute random variables).
   (a) Find $P(1/2 \le X \le 3/4)$.
   (b) What is the probability that at least two of the three independent observations are greater than $1/2$?

5. The joint probability distribution for random variables X and Y is given by
   $f(x, y) = k, x = 0, 1, 2, \cdots$ and $y = 0, 1, 2, \cdots 3 - x$.
   (a) Find k.
   (b) Find the marginal distributions for X and Y.

6. Use the central limit theorem to approximate the probability that the sum is 34 when 12 dice are thrown.

7. The maximum weight an elevator can carry is 1600 lb. If the weights of individuals using the elevator are $N(150, 10)$, what is the probability that the elevator will be overloaded?

8. A continuous probability distribution is defined as

$$f(x) = \begin{cases} x, & 0 < x < 1 \\ k, & 1 < x < 2 \\ k(3 - x), & 2 < x < 3 \end{cases}$$

   Find k.

Chapter 9

Statistical Inference I

CHAPTER OBJECTIVES:

- to study *statistical inference*: how conclusions can be drawn from samples
- to study both point and interval estimation
- to learn about two types of errors in hypothesis testing
- to study operating characteristic curves and the influence of sample size on our conclusions.

We now often encounter the results of sample surveys. We read about political polls on how we feel about various issues; television networks and newspapers conduct surveys about the popularity of politicians and how elections are likely to turn out. These surveys are normally performed with what might be thought as a relatively small sample of the population being surveyed. These sample sizes are in reality quite adequate for drawing inferences or conclusions; it all depends on how accurate one wishes the survey to be.

How is it that a sample from a population can give us information about that population? After all, some samples may be quite typical of the population from which they are chosen, while other samples may be very unrepresentative of the population from which they are chosen. This heavily depends on the theory of probability that we have developed. We explore some ideas here and explain some of the basis for *statistical inference*: the process of drawing conclusions from samples.

Statistical inference is usually divided into two parts: *estimation* and *hypothesis testing*. We consider each of these topics now.

A Probability and Statistics Companion, John J. Kinney
Copyright © 2009 by John Wiley & Sons, Inc.

ESTIMATION

Assuming we do not already know the answer, suppose we wish to guess the age of your favorite teacher. Most people would give an exact response: 52, 61, 48, and so on. These are *estimates* of the age and, since they are exact, are called *point estimates.*

Now suppose it is very important that our response be correct. We are unlikely to estimate the correct age exactly, so a point estimate may not be a good response. How else can one respond to the question? Perhaps a better response is, "I think the teacher is between 45 and 60." We might feel, in some sense, that the interval might have a better chance of being correct, that is, in containing the true age, than a point estimate. But we must be very careful in interpreting this, as we will see.

The response in our example is probably based on observation and does not involve a random sample, so we cannot assign any probability or likelihood to the interval. We now turn to the situation where we have a random sample and wish to create an interval estimate. Such intervals are called confidence intervals.

CONFIDENCE INTERVALS

EXAMPLE 9.1   *A Confidence Interval*

Consider a normal random variable X, whose variance σ^2 is known but whose mean μ is unknown. The central limit theorem tells us that the random variable representing the mean of a sample of n random observations, $\overline{X}$, has a $N(\mu, \sigma/\sqrt{n})$ distribution.

So, one could say that

$$P\left(-1.645 \leq \frac{\overline{X} - \mu}{\frac{\sigma}{\sqrt{n}}} \leq 1.645\right) \doteq 0.90$$

This is true because $(\overline{X} - \mu)/(\sigma/\sqrt{n})$ is a normally distributed random variable. We could have chosen many other true statements such as

$$P\left(-1.96 \leq \frac{\overline{X} - \mu}{\frac{\sigma}{\sqrt{n}}} \leq 1.96\right) \doteq 0.95$$

or

$$P\left(-1.282 \leq \frac{\overline{X} - \mu}{\frac{\sigma}{\sqrt{n}}}\right) \doteq 0.90$$

We have, of course, an infinite number of choices for this normal random variable. The probabilities in the statements above are called *confidence coefficients*. There is an infinity of choices for the confidence coefficient. Once a confidence coefficient is selected and a symmetric interval is decided, the normal z-values can be found by using computer or hand-held calculator or from published tables. The probability statements are all based on the fact that $(\overline{X} - \mu)/(\sigma/\sqrt{n})$ is a legitimate random variable that will vary from sample to sample.

Now rearrange the inequality in the statement

$$P\left(-1.645 \leq \frac{\overline{X} - \mu}{\frac{\sigma}{\sqrt{n}}} \leq 1.645\right) \doteq 0.90$$

to read

$$-1.645 \cdot \frac{\sigma}{\sqrt{n}} \leq \overline{X} - \mu \leq 1.645 \cdot \frac{\sigma}{\sqrt{n}}$$

We see further that

$$-1.645 \cdot \frac{\sigma}{\sqrt{n}} - \overline{X} \leq -\mu \leq 1.645 \cdot \frac{\sigma}{\sqrt{n}} - \overline{X}$$

Now if we solve for μ and rearrange the inequalities, we find that

$$\overline{X} - 1.645 \cdot \frac{\sigma}{\sqrt{n}} \leq \mu \leq \overline{X} + 1.645 \cdot \frac{\sigma}{\sqrt{n}}$$

Note now that the end points of the inequality above are both known, since $\overline{X}$ and n are known from the sample and we presume that σ is known.

The interval we calculated above is called a 95% *confidence interval*.

We have omitted these statements as probability statements. The reason for this is that the statement

$$P\left(-1.645 \leq \frac{\overline{X} - \mu}{\frac{\sigma}{\sqrt{n}}} \leq 1.645\right) \doteq 0.90$$

is a statement about a random variable and is a legitimate probability statement. The result

$$\overline{X} - 1.645 \cdot \frac{\sigma}{\sqrt{n}} \leq \mu \leq \overline{X} + 1.645 \cdot \frac{\sigma}{\sqrt{n}}$$

however, is not a probability statement about μ. Although μ is unknown, it is a constant and so it is either in the interval from $\overline{X} - 1.645 \cdot \sigma/\sqrt{n}$ to $\overline{X} + 1.645 \cdot \sigma/\sqrt{n}$ or it is not. So the probability that μ is in the interval is either 0 or 1. What meaning, then, are we to give to the 0.90 with which we began?

We interpret the final result in this way: 90% of all possible intervals calculated will contain the unknown, constant value μ, and 10% of all possible intervals calculated will not contain the unknown, constant value μ.   ∎

EXAMPLE 9.2   *Several Confidence Intervals*

To illustrate the ideas in Example 9.1, we drew 20 samples of size 3 from a normal distribution with mean 2 and standard deviation 2. The sample means are then approximately normally distributed with mean 2 and standard deviation $2/\sqrt{3} = 1.1547$.

A graph of the 20 confidence intervals generated is shown in Figure 9.1.

As it happened, exactly 19 of the 20 confidence intervals contain the mean, indicated by the vertical line. This occurrence is not always to be expected.   ∎

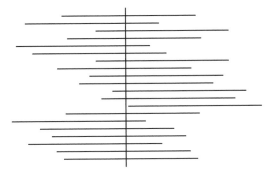

Figure 9.1   Confidence intervals.

We have considered the estimation of some unknown mean μ by using a random sample and constructing a confidence interval. We will consider other confidence intervals subsequently, but now we turn to the other major part of statistical inference, *hypothesis testing*.

HYPOTHESIS TESTING

If we choose a random sample from a probability distribution and calculate the sample mean, $\overline{X} = 32.4$, can we confidently believe that the sample was chosen from a population with $\mu = 26$? The answer of course lies in both the variability of $\overline{X}$ and the confidence we need to place in our conclusion.

Two examples will be discussed here, the first from a discrete distribution and the second from a continuous distribution. We confess now that each of the examples is somewhat artificial and is used primarily to introduce some ideas so that they can be brought out clearly. It is crucial that these ideas be clearly understood before we proceed to more realistic problems.

EXAMPLE 9.3   *Germinating Bulbs*

A horticulturist is experimenting with an altered bulb for a large plant. From previous experience, she knows that the percentage of these bulbs that germinate is either 50% or 75%. To decide which germination rate is correct, she plans an experiment involving 15 of these altered bulbs and records the number of bulbs that germinate.

We assume that the number of bulbs that germinate follows a binomial model, that is, a bulb either germinates or it does not, the bulbs behave independently, and the probability of germination is constant. If in fact the probability is 50% that a bulb germinates and if X is the random variable denoting the number of bulbs that germinate, then

$$P(X = x) = \binom{15}{x}(0.50)^x \cdot (0.50)^{15-x} \quad \text{for } x = 0, 1, 2, \cdots, 15$$

while if the probability is 75% that a bulb germinates, then

$$P(X = x) = \binom{15}{x}(0.75)^x \cdot (0.25)^{15-x} \quad \text{for } x = 0, 1, 2, \cdots, 15$$

We should first consider the probabilities of all the possible outcomes from the experiment. These are shown in Table 9.1.

Table 9.1   Probabilities for Example 9.3

x	$p = 0.50$	$p = 0.75$
0	0.0000	0.0000
1	0.0004	0.0000
2	0.0032	0.0000
3	0.0139	0.0000
4	0.0417	0.0001
5	0.0916	0.0007
6	0.1527	0.0034
7	0.1964	0.0131
8	0.1964	0.0393
9	0.1527	0.0917
10	0.0916	0.1651
11	0.0417	0.2252
12	0.0139	0.2252
13	0.0032	0.1559
14	0.0004	0.0668
15	0.0000	0.0134

The statements that 50% of the bulbs germinate or 75% of the bulbs germinate are called *hypotheses*. They are *conjectures* about the behavior of the bulbs. We will formalize these hypotheses as

$$H_0 : p = 0.50$$
$$H_a : p = 0.75$$

We have called $H_0 : p = 0.50$ the *null hypothesis* and $H_a : p = 0.75$ the *alternative hypothesis*. Now we must decide between them. If we decide that the null hypothesis is correct, then we *accept* the null hypothesis and *reject* the alternative hypothesis. On the contrary, if we reject the null hypothesis then we accept the alternative hypothesis How should we decide? The decision process is called *hypothesis testing*.

In this case, we would certainly look at the number of bulbs that germinate. If in fact 75% of the bulbs germinate, then we would expect a large number of the bulbs to germinate.

It would appear, if a large number of bulbs germinate, say 11 or more, that we would then reject the null hypothesis (that $p = 0.50$) and accept the alternative hypothesis (that $p = 0.75$).

In coming to this test, we cannot reach a decision with certainty because our conclusion is based on a sample, a small one at that in this instance. What are the *risks* involved? There are two risks or errors that we can make: we could reject the null hypothesis when it is actually true or we could accept the null hypothesis when it is false. Let us consider each of these.

Rejecting the null hypothesis when it is true is called a Type I error. In this case, we reject the null hypothesis when the number of germinating bulbs is 11 or more. The probability this occurs when the null hypothesis is true is

$$P(\text{Type I error}) = 0.0417 + 0.0139 + 0.0032 + 0.0004 + 0.0000 = 0.0592$$

So about 6% of the time, bulbs that have a germination rate of 50% will behave as if the germination rate were 75%.

Accepting the null hypothesis when it is false is called a Type II error. In this case, we accept the null hypothesis when the number of germinating bulbs is 10 or less. The probability this occurs when the null hypothesis is false is

$$P(\text{Type II error}) = 0.0000 + 0.0000 + 0.0000 + 0.0000 + 0.0001 + 0.0007$$
$$+ 0.0033 + 0.0131 + 0.0393 + 0.0917 + 0.1651$$
$$= 0.3133$$

So about 31% of the time, bulbs with a germination rate of 75% will behave as though the germination rate were only 50%.

The experiment will always result in some value of X. We must decide in advance which values of X cause us to accept the null hypothesis and which values of X cause us to reject the null hypothesis.

The values of X that cause us to reject H_0 comprise what we call the *critical region* for the test.

In this case, large values of X are more likely to come from a distribution with $p = 0.75$ than from a distribution with $p = 0.50$. We have used the critical region $X \geq 11$ here.

So it is reasonable to conclude that if $X \geq 11$, then $p = 0.75$.

The errors calculated above are usually denoted by α and β. In general then

$$\alpha = P(H_0 \text{ is rejected if it is true})$$

where α is often called the *size* or the *significance level* of the test.

The size of the Type II error is denoted by β. In general then

$$\beta = P(H_0 \text{ is accepted if it is false}).$$

In this case, with the critical region $X \geq 11$, we find $\alpha = 0.0542$ and $\beta = 0.3133$.

Note that α and β are calculated under quite different assumptions, since α presumes the null hypothesis true and β presumes the null hypothesis false, so they bear no particular relationship to one another.

It is of course possible to decrease α by reducing the critical region to say $X \geq 12$. This produces $\alpha = 0.0175$, but unfortunately, the Type II error increases to 0.5385. The only way to decrease both α and β simultaneously is to increase the sample size.

Finally note that both α and β increase or decrease in finite amounts. It is not possible to find a critical region that would produce α between the values 0.0175 and 0.0542. It is possible to decrease both α and β by increasing the sample size as we now show. ∎

EXAMPLE 9.4   *Increasing the Sample Size*

Suppose now that the experimenter in the previous example has 100 bulbs with which to experiment. We could work out all the probabilities for the 101 possible values for X and would most certainly use a computer to do this. For this sample size, the binomial distribution is very normal-like with a maximum at the expected value, np, and standard deviation $\sqrt{n \cdot p \cdot (1 - p)}$. In either case here, $n = 100$. If $p = 0.50$, we find the probability distribution centered about $np = 100 \cdot 0.50 = 50$ with standard deviation $\sqrt{100 \cdot 0.50 \cdot 0.50} = 5$. Since we are seeking a critical region in the upper tail of this distribution, we look at values of X at least one standard deviation from the mean, so we start at $X = 55$. We show some probabilities in Table 9.2.

Table 9.2

Critical region	α	β
$X \geq 56$	0.1356	0.0000
$X \geq 57$	0.0967	0.0000
$X \geq 58$	0.0443	0.0001
$X \geq 59$	0.0284	0.0001
$X \geq 60$	0.0176	0.0003
$X \geq 61$	0.0105	0.0007
$X \geq 62$	0.0060	0.0014

We see that α decreases as we move to the right on the probability distribution and that β increases. We have suggested various critical regions here and determined the resulting values of the errors. This raises the possibility that the size of one of the errors, say α, is chosen in advance and then a critical region found that produces this value of α. The consequence of this is shown in Table 9.2. It is not possible, for example, to choose $\alpha = 0.05$ and find an appropriate critical region. This is because the random variable is discrete in this case. If the random variable were continuous, then it is possible to specify α in advance. We show how this is so in the next example.   ■

EXAMPLE 9.5   *Breaking Strength*

The breaking strength of steel wires used in elevator cables is a crucial characteristic of these cables. The cables can be assumed to come from a population with known $\sigma = 400$ lb. Before accepting a shipment of these steel wires, an engineer wants to be confident that $\mu > 10{,}000$ lb. A sample of 16 wires is selected and their mean breaking strength $\overline{X}$ is measured.

It would appear sensible to test the null hypothesis $H_0 : \mu = 10{,}000$ lb against the alternative $H_a : \mu < 10{,}000$. A test will be based on the sample mean, $\overline{X}$. The central limit theorem tells us that

$$\overline{X} \sim N\left(\mu, \frac{\sigma}{\sqrt{n}}\right)$$

In this case, we have

$$\overline{X} \sim N\left(\mu, \frac{400}{\sqrt{16}}\right) = N(\mu, 100)$$

If the critical region has size 0.05, so that $\alpha = 0.05$, then we would select a critical region in the left tail of the normal curve. The situation is shown in Figure 9.2.

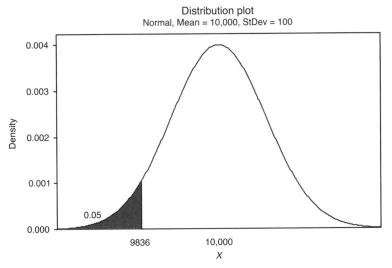

Figure 9.2

The value of shaded area in Figure 9.2 is 0.05, so the z−score is -1.645. This means that the critical value of $\overline{X}$ is $-1.645 = (\overline{x} - 10{,}000)/(100)$ or $\overline{x} = 10000 - 164.5 = 9835.5$. So the null hypothesis should be rejected if the sample mean is less than 9835.5 lb.

β AND THE POWER OF A TEST

What is β, the size of the Type II error in this example? We recall that

$$\beta = P(H_0 \text{ is accepted if it is false})$$

or

$$\beta = P(H_0 \text{ is accepted if the alternative is true})$$

We could calculate β easily in our first example since in that case we had a specific alternative to deal with (namely, $p = 0.75$). However, in this case, we have an infinity of alternatives ($\mu < 10{,}000$) to deal with.

The size of β depends upon which of these specific alternatives is chosen. We will show some examples. We use the notation β_{alt} to denote the value of β when a particular alternative is selected.

First, consider

$$\beta_{9800} = P(\overline{X} > 9835.5 \quad \text{if } \mu = 9800)$$

$$= P\left(Z > \frac{9835.5 - 9800}{100} = 0.355\right) = 0.361295$$

So for this test the probability we accept $H_0 : \mu = 10{,}000$ if in fact $\mu = 9800$ is over 36%. Now let us try some other values for the alternative.

$$\beta_{9900} = P(\overline{X} > 9835.5 \quad \text{if } \mu = 9900)$$

$$= P\left(Z > \frac{9835.5 - 9900}{100} = -0.645\right) = 0.740536$$

So almost 3/4 of the time this test will accept $H_0 : \mu = 10{,}000$ if in fact $\mu = 9900$. β then highly depends upon the alternative hypothesis.

We can in fact show this dependence in Figure 9.3 in a curve that plots β_{alt} against the specific alternative.

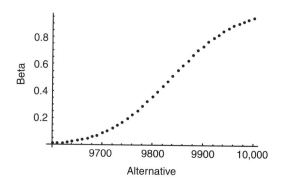

Figure 9.3

This curve is often called the *operating characteristic curve* for the test.

This curve is not part of a normal curve; in fact, it has no algebraic equation, each point on it being calculated in the same manner as we have done in the previous two examples.

We know that $\beta_{alt} = P(\text{accept } H_0 \text{ if it is false})$ so $1 - \beta_{alt} = P(\text{reject } H_0 \text{ if it is false})$. This is the probability that the null hypothesis is correctly rejected. $1 - \beta_{alt}$ is called the *power of the test* for a specific alternative.

A graph of the power of the test is shown in Figure 9.4. Figure 9.4 also shows the power of the test if the sample size were to be increased from 16 to 100. The graph indicates that the sample of 100 is more likely to reject a false H_0 than is the sample of 16. ■

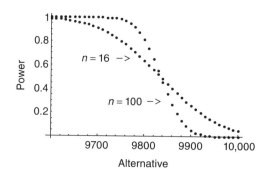

Figure 9.4

p-VALUE FOR A TEST

We have discussed selecting a critical region for a test in advance and the disadvantages of proceeding in that way. We abandoned that approach for what would appear to be a more reasonable one, that is, selecting α in advance and calculating the critical region that results. Selecting α in advance puts a great burden upon the experimenter. How is the experimenter to know what value of α to choose? Should 5% or 6% or 10% or 22% be selected? The choice often depends upon the sensitivity of the experiment itself. If the experiment involves a drug to combat a disease, then α should be very small; however, if the experiment involves a component of a nonessential mechanical device, then the experimenter might tolerate a somewhat larger value of α. Now we abandon that approach as well. But then we have a new problem: if we do not have a critical region and if we do not have α either, then how can we proceed?

Suppose, to be specific, that we are testing $H_0 : \mu = 22$ against $H_1 : \mu \neq 22$ with a sample of $n = 25$ and we know that $\sigma = 5$. The experimenter reports that the observed $\overline{X} = 23.72$. We could calculate that a sample of 25 would give this result, or a result greater than 23.72, if the true mean were 22. This is found to be

$$P(\overline{X} \geq 23.72 \quad \text{if } \mu = 22)$$
$$= P(Z \geq 1.72)$$
$$= 0.0427162$$

Since the test is two sided, the phrase "a result more extreme" is interpreted to mean $P(|Z| \geq 1.72) = 2 \cdot 0.0427162 = 0.08\,5432$. This is called the *p-value for the test*.

This allows the experimenter to make the final decision, either to accept or reject the null hypothesis, depending entirely upon the size of this probability. If the *p*-value is very large, one would normally accept the null hypothesis, while if it is very small, one would know that the result is in one of the extremities of the distribution and reject the null hypothesis. The decision of course is up to the experimenter.

Here is a set of rules regarding the calculation of the *p*-value. We assume that z is the observed value of Z and that the null hypothesis is $H_0 : \mu = \mu_0$:

Alternative	*p*-value
$\mu > \mu_0$	$P(Z > z)$
$\mu < \mu_0$	$P(Z < z)$
$\mu \neq \mu_0$	$P(|Z| > z)$

The *p*-values have become popular because they can be easily computed. Tables offer great limitations and their use generally allows only approximations to *p*-values. Statistical computer programs commonly calculate *p*-values.

CONCLUSIONS

We have introduced some basic ideas regarding statistical inference—the process by which we draw inferences from samples—using confidence intervals or hypothesis testing.

   We will continue our discussion of statistical inference in the next chapter where we look at the situation when nothing is known about the population being sampled. We will also learn about comparing two samples drawn from possibly different populations.

EXPLORATIONS

1. Use a computer to select 100 samples of size 5 each from a $N(0, 1)$ distribution. Compute the mean of each sample and then find how many of these are within the interval -0.73567 and 0.73567. How many means are expected to be in this interval?

2. In the past, 20% of the production of a sensitive component is unsatisfactory. A sample of 20 components shows 6 items that must be reworked. The manufacturer is concerned that the percentage of components that must be reworked has increased to 30%.
   (a) Form appropriate null and alternative hypotheses.
   (b) Let X denote the number of items in a sample that must be reworked. If the critical region is $\{x \mid x \geq 9\}$, find the sizes of both Type I and Type II errors.
   (c) Choose other critical regions and discuss the implications of these on both types of errors.

Chapter 10

Statistical Inference II: Continuous Probability Distributions II—Comparing Two Samples

CHAPTER OBJECTIVES:

- to test hypotheses on the population variance
- to expand our study of statistical inference to include hypothesis tests for a mean when the population standard deviation is not known
- to test hypotheses on two variances
- to compare samples from two distributions
- to introduce the Student t, Chi-Squared, and F probability distributions
- to show some applications of the above topics.

In the previous chapter, we studied testing hypotheses on a single mean, but we presumed, while we did not know the population mean, that we did know the standard deviation. Now we examine the situation where nothing is known about the sampled population. First, we look at the population variance.

THE Chi-SQUARED DISTRIBUTION

We have studied the probability distribution of the sample mean through the central limit theorem. To expand this inference when the population standard deviation σ is

A Probability and Statistics Companion, John J. Kinney
Copyright © 2009 by John Wiley & Sons, Inc.

unknown, we must be able to make inferences about the population variance. We begin with a specific example.

One hundred samples of size 5 were selected from a $N(0, 1)$ distribution and on each case the sample variance s^2 was calculated. In general,

$$s^2 = \frac{\sum\limits_{i=1}^{n}(x_i - \bar{x}_i)^2}{n - 1}$$

This can be shown to be

$$s^2 = \frac{n \sum\limits_{i=1}^{n} x_i^2 - \left(\sum\limits_{i=1}^{n} x_i\right)^2}{n(n - 1)}$$

In this case, $n = 5$.

Here are some of the samples, the sample mean, and the sample variance:

Sample	$\bar{x}$	s^2
−0.815117, −0.377894, 1.12863, −0.550026, −0.969271	−0.316736	0.705356
0.172919, −0.989559, 0.448382, 0.778734, −1.37236	−0.192377	0.878731
.	.	.
.	.	.
0.0723954, 0.633443, −0.437318, −0.684865, −1.33975	−0.351219	0.561233

A graph of the values of s^2 is shown in Figure 10.1.

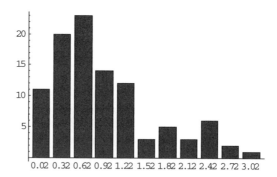

Figure 10.1

It is clear that the graph has a somewhat long right-hand tail and of course must be nonnegative. It is certainly not normal due to these facts. Our sample showed a mean value for the sample variances to be 1.044452, and the variance of these variances was 0.521731.

Now we state, without proof, the following theorem:

Theorem 10.1   If samples are selected from a $N(\mu, \sigma)$ distribution and the sample variance s^2 is calculated for each sample, then $(n - 1)s^2/\sigma^2$ follows the chi-squared probability distribution with $n - 1$ degrees of freedom (χ_{n-1}^2).

Also

$$E(s^2) = \sigma^2 \text{ and } \text{Var}(s^2) = \frac{2\sigma^4}{n-1}$$

The proof of this theorem can be found in most texts on mathematical statistics. The theorem says that the probability distribution depends on the sample size n and that the distribution has a parameter $n-1$, which we call the *degrees of freedom*.

We would then expect, since $\sigma^2 = 1$, that $E(s^2) = 1$; our sample showed a mean value of the variances to be 1.044452. We would also expect the variance of the sample variances to be $2\sigma^4/(n-1) = 2 \cdot 1/4 = 1/2$; our variances showed a variance of 0.521731, so our samples tend to agree with the results stated in the theorem.

Now we show in Figure 10.2, the chi-squared distribution with $n-1 = 4$ degrees of freedom (which we abbreviate as χ_4^2).

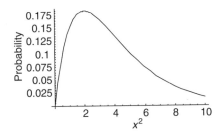

Figure 10.2

Finally, in Figure 10.3, we superimpose the χ_4^2 distribution on our histogram of sample variances.

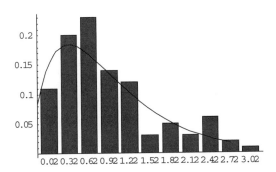

Figure 10.3

Note that exponent 2 carries no meaning whatsoever; it is simply a symbol and alerts us to the fact that the random variable is nonnegative. It is possible, but probably useless, to find the probability distribution for χ, as we could find the probability distribution for the square root of a normal random variable.

We close this section with an admonition: while the central limit theorem affords us the luxury of sampling from any probability distribution, the probability distribution of χ^2 highly depends on the fact that the samples arise from a normal distribution.

STATISTICAL INFERENCE ON THE VARIANCE

While the mean value of the diameter of a manufactured part is very important for the part to fit into a mechanism, the variance of the diameter is also crucial so that parts do not vary widely from their target value. A sample of 12 parts showed a sample variance $s^2 = 0.0025$. Is this the evidence that the true variance σ^2 exceeds 0.0010?

To solve the problem, we must calculate some probabilities.

One difficulty with the chi-squared distribution, and indeed with almost all practical continuous probability distributions, is the fact that areas, or probabilities, are very difficult to compute and so we rely on computers to do that work for us. The computer system Mathematica and the statistical program Minitab® have both been used in this book for these calculations and the production of graphs.

Here are some examples where we need some points on χ^2_{11}:

We find that $P(\chi^2_{11} < 4.57) = 0.05$ and so $P((n-1)s^2/\sigma^2 < 4.57) = 0.05$, which means that $P(\sigma^2 > (n-1)s^2/4.57) = 0.05$ and in this case we find $P(\sigma^2 > 11 \cdot 0.0025/4.57) = P(\sigma^2 > 0.0060175) = 0.05$ and so we have a confidence interval for σ^2.

Hypothesis tests are carried out in a similar manner. In this case, as in many other industrial examples, we are concerned that the variance may be too large; small variances are of course desirable. For example, consider the hypotheses from the previous example, $H_0 : \sigma^2 = 0.0010$ and $H_A : \sigma^2 > 0.0010$.

From our data, where $s^2 = 0.0025$, and from the example above, we see that this value for s^2 is in the rejection region. We also find that $\chi^2_{11} = (n-1)s^2/\sigma^2 = 11 \cdot 0.0025/0.0010 = 27.5$ and we can calculate that $P(\chi^2_{11} > 27.5) = 0.00385934$, and so we have a p-value for the test.

Without a computer this could only be done with absolutely accurate tables. The situation is shown in Figure 10.4.

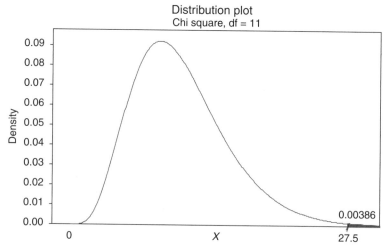

Figure 10.4

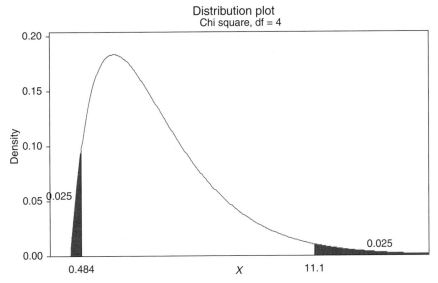

Figure 10.5

Now we look at some graphs with other degrees of freedom in Figures 10.5 and 10.6.

The chi-squared distribution becomes more and more "normal-like" as the degrees of freedom increase. Figure 10.7 shows a graph of the distribution with 30 degrees of freedom.

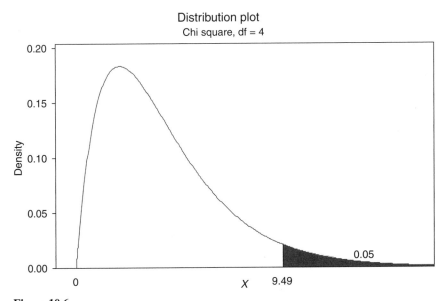

Figure 10.6

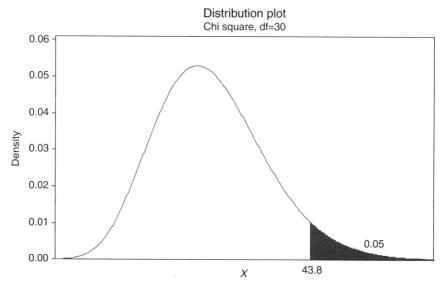

Figure 10.7

We see that $P(\chi^2_{30} > 43.8) = 0.05$. It can be shown that $E(\chi^2_n) = n$ and that $\mathrm{Var}(\chi^2_n) = 2n$. If we approximate χ^2_{30} by a normal distribution with mean 30 and standard deviation $\sqrt{60} = 7.746$, we find that the point with 5% of the curve in the right-hand tail is $30 + 1.645\sqrt{60} = 42.7$, so the approximation is not too bad. The approximation is not very good, however, for small degrees of freedom.

Now we are able to consider inferences about the sample mean when σ is unknown.

STUDENT *t* DISTRIBUTION

In the previous chapter, we have used the central limit theorem to calculate confidence intervals and to carry out tests of hypotheses. In noting that the random variable $(\bar{x} - \mu)/(\sigma/\sqrt{n})$ follows a $N(0, 1)$ distribution, we rely heavily on the fact that the standard deviation of the population, σ, is known. However, in many practical situations, both population parameters, μ and σ, are unknown. In 1908, W.G. Gossett, writing under the pseudonym "Student" discovered the following:

Theorem 10.2   The ratio of a $N(0, 1)$ random variable divided by the square root of a chi-squared random variable divided by its degrees of freedom (say n) follows a Student t distribution with n degrees of freedom (t_n).

Since $(\bar{x} - \mu)/(\sigma/\sqrt{n})$ is a $N(0, 1)$ random variable, and $(n - 1)s^2/\sigma^2$ is a χ^2_{n-1} variable, it follows that $[(\bar{x} - \mu)/(\sigma/\sqrt{n})]/\sqrt{(n - 1)s^2/\sigma^2(n - 1)}$ follows a t_{n-1}

probability distribution. But

$$\frac{\dfrac{\bar{x} - \mu}{\sigma/\sqrt{n}}}{\sqrt{\dfrac{(n-1)s^2}{\sigma^2(n-1)}}} = \frac{\bar{x} - \mu}{s/\sqrt{n}}$$

So we have a random variable that involves μ alone and values calculated from a sample. It is far too simple to say that we simply replace σ by s in the central limit theorem; much more than that has transpired. The sampling can arise from virtually any distribution due to the fact that the central limit theorem is almost immune to the underlying distribution.

We show two typical Student t distributions in Figure 10.8 (which was produced by Minitab).

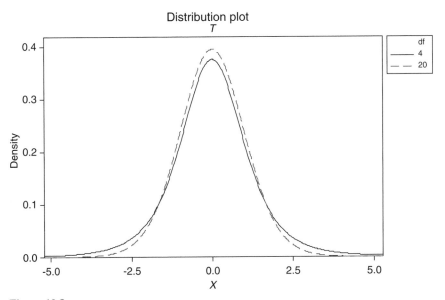

Figure 10.8

The curves have 4 and 20 degrees of freedom, respectively, and each appears to be "normal-like." Here is a table of values of v so that the probability the table value exceeds v is 0.05. The number of degrees of freedom is n.

n	v
5	2.02
10	1.81
20	1.72
30	1.70
40	1.68

This is some evidence that the 0.05 point approaches that of the $N(0, 1)$ distribution, 1.645, but the degrees of freedom, depending on the sample size, remain crucial.

Again a computer is essential in doing calculations as the following example shows.

EXAMPLE 10.1   *A Sample*

A sample of size 14 showed that $\bar{x} = 29.8$ and $s^2 = 123$. Find the p-value for the test $H_0 : \mu = 26$ and $H_a : \mu \neq 26$.

Here we find that $t_{13} = \frac{29.8 - 26}{\sqrt{123/14}} = 4.797$ and $P(t_{13} > 4.797) = 1.7439 \cdot 10^{-4}$, a rare event indeed.   ∎

TESTING THE RATIO OF VARIANCES: THE *F* DISTRIBUTION

So far we have considered inferences from single samples, but we will soon turn to comparing two samples, possibly arising from two different populations. So we now investigate comparing variances from two different samples. We need the following theorem whose proof can be found in texts on mathematical statistics.

Theroem 10.3   The ratio of two independent chi-squared random variables, divided by their degrees of freedom, say n in the numerator and m in the denominator, follows the F distribution with n and m degrees of freedom.

We know from a previous section that $(n - 1)s^2/\sigma^2$ is a χ^2_{n-1} random variable, so if we have two samples, with degrees of freedom n and m, then using the theorem,

$$\frac{\dfrac{ns_1^2}{\sigma_1^2 n}}{\dfrac{ms_2^2}{\sigma_2^2 m}} = \frac{\dfrac{s_1^2}{\sigma_1^2}}{\dfrac{s_2^2}{\sigma_2^2}} = F[n, m]$$

Figure 10.9 shows a graph of a typical F distribution, this with 7 and 9 degrees of freedom.

The graph also shows the upper and lower 2.5% points.

Now notice that the reciprocal of an F random variable is also an F random variable but with the degrees of freedom interchanged, so $\frac{1}{F[n, m]} = F[m, n]$. This fact can be used in finding critical values. Suppose $P[F[n, m] < v] = \alpha$. This is equivalent to $P[1/F[n, m] > 1/v]$ or $P[F[m, n] > 1/v] = \alpha$. So the reciprocal of the lower α point on $F[n, m]$ is the upper α point on the $F[m, n]$ distribution.

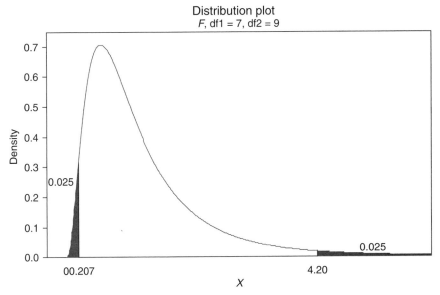

Figure 10.9

EXAMPLE 10.2   *Two Samples*

Independent samples from two normal distributions gave the following data: $n_1 = 22$, $s_1^2 = 100$, $n_2 = 13$, and $s_2^2 = 200$. We seek a 90% confidence interval for the ratio of the true variances, σ_1^2/σ_2^2.

We know that

$$\frac{\frac{s_2^2}{\sigma_2^2}}{\frac{s_1^2}{\sigma_1^2}} = \frac{200}{\sigma_2^2} \cdot \frac{\sigma_1^2}{100} = \frac{2\sigma_1^2}{\sigma_2^2} = F[12, 21]$$

so we find, from Figure 10.10, that
$P[0.395 < 2\sigma_1^2/\sigma_2^2 < 2.25] = 0.90$, which can also be written as

$$P\left[0.1975 < \frac{\sigma_1^2}{\sigma_2^2} < 1.125\right] = 0.90$$

It is interesting to note, while one sample variance is twice the other, that the confidence interval contains 1 so the hypothesis $H_0 : \sigma_1^2 = \sigma_2^2$ against $H_a : \sigma_1^2 \neq \sigma_2^2$ would be accepted with $\alpha = 0.10$.

This is an indication of the great variability of the ratio of variances. ∎

We now turn to other hypotheses involving two samples.

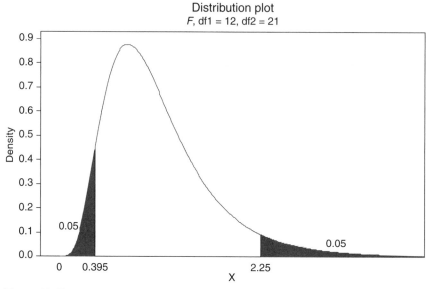

Figure 10.10

TESTS ON MEANS FROM TWO SAMPLES

It is often necessary to compare two samples with each other. We may wish to compare methods of formulating a product; different versions of standardized tests may be compared to decide if they in fact test equally able candidates equally; production methods may be compared with respect to the quality of the products each method produces. Two samples may be compared by comparing a variety of sample statistics; we consider comparing only means and variances here.

EXAMPLE 10.3   *Two Production Lines*

Two production lines, called X and Y, are making specialized heaters. Samples are selected from each production line, the thermostats set at $140°$, and then the actual temperature in the heater is measured. The results of the sampling are given in Table 10.1.

Graphs should be drawn from the data so that we may make an initial visual inspection. Comparative dot plots of the samples are shown in Figure 10.11.

The data from the production line X appear to be much more variable than that from production line Y. It also appears that the samples are centered about different points, so we calculate some sample statistics and we find that

$$n_x = 15, \bar{x} = 138.14, s_x = 6.95$$

$$n_y = 25, \bar{y} = 144.02, s_y = 3.15$$

Table 10.1

X	Y
147.224	135.648
121.482	140.083
142.691	140.970
127.155	138.990
147.766	148.490
131.562	145.757
139.844	145.740
139.585	146.324
142.966	145.472
140.058	145.332
139.553	140.822
137.973	145.022
137.343	145.496
137.151	147.103
139.809	144.753
	145.598
	145.471
	145.319
	143.103
	145.676
	141.644
	144.381
	146.797
	139.065
	147.352

Comparative dot plots for Example 10.3

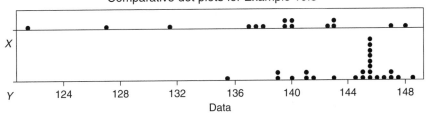

Figure 10.11

   The real question here is how these statistics would vary were we to select large numbers of pairs of samples. The central limit theorem can be used if assumptions can be made concerning the population variances.

   We formulate the problem as follows. We wish to test the hypothesis $H_0 : \mu_X = \mu_Y$ against $H_1 : \mu_X \neq \mu_Y$.

   We know that $E[\bar{X} - \bar{Y}] = \mu_X - \mu_Y$ and that

$$\text{Var}[\bar{X} - \bar{Y}] = \frac{\sigma_X^2}{n_x} + \frac{\sigma_Y^2}{n_y}$$

and that each of the variables $\bar{X}$ and $\bar{Y}$ is individually approximately normally distributed by the central limit theorem. It can be shown that the difference between normal variables is also normal so

$$z = \frac{(\bar{X} - \bar{Y}) - (\mu_X - \mu_Y)}{\sqrt{\frac{\sigma_X^2}{n_x} + \frac{\sigma_Y^2}{n_y}}}$$

is a $N(0, 1)$ variable.

The statistic z can be used to test hypotheses or to construct confidence intervals if the variances are known. Consider for the moment that we can assume that the populations have equal variances, say $\sigma_X^2 = \sigma_Y^2 = 30.0$. Then

$$z = \frac{(138.14 - 144.02) - 0}{\sqrt{\frac{30}{15} + \frac{30}{25}}} = -3.287$$

The p-value for this test is approximately 0.001, so the null hypothesis would most probably not be accepted.

We could also use z to construct a confidence interval. Here

$$P\left[(\bar{X} - \bar{Y}) - 1.96\sqrt{\frac{\sigma_X^2}{n_x} + \frac{\sigma_Y^2}{n_y}} \leq \mu_X - \mu_Y \leq (\bar{X} - \bar{Y}) + 1.96\sqrt{\frac{\sigma_X^2}{n_x} + \frac{\sigma_Y^2}{n_y}}\right] = 0.95$$

which becomes in this case the interval from -9.3862 to -2.3739. Since 0 is not in this interval, the hypothesis of equal means is rejected. ∎

Larger samples reduce the width of the confidence interval.

Knowledge of the population variances in the previous example may be regarded as artificial or unusual, although it is not infrequent, when data have been gathered over a long period of time, that some idea of the size of the variance is known. Now we consider the case where the population variances are unknown. There are two cases: the unknown variances are equal or they are not equal. We give examples of the procedure in each case.

EXAMPLE 10.4 *Using Pooled Variances*

If the population variances are known to be equal, with true value σ^2, then we form an estimate of this common value, which we denote by s_p^2, where

$$s_p^2 = \frac{(n_x - 1)s_X^2 + (n_y - 1)s_Y^2}{n_x + n_y - 2}$$

Here s_p^2 is often called the *pooled variance*. The sampling must now be done from normal distributions.

We replace each of the unknown, but equal, variances with the pooled variance. Then it is known that

$$t_{n_X + n_Y - 2} = \frac{(\bar{X} - \bar{Y}) - (\mu_X - \mu_Y)}{s_p\sqrt{\frac{1}{n_X} + \frac{1}{n_Y}}}$$

In this case, we find that

$$s_p^2 = \frac{14 \cdot 6.95^2 + 24 \cdot 3.15^2}{15 + 25 - 2} = 24.0625$$

and so

$$t_{38} = \frac{(138.14 - 144.02) - 0}{\sqrt{24.0625}\sqrt{\frac{1}{15} + \frac{1}{25}}} = -3.670$$

The p-value for the test is then about 0.0007 leading to the rejection of the hypothesis that the population means are equal. ∎

EXAMPLE 10.5   *Unequal Variances*

Now we must consider the case where the population variances are unknown and cannot be presumed to be equal. Regrettably, we do not know the exact probability distribution of any statistic involving the sample data in this case. This unsolved problem is known as the Behrens–Fisher problem; several approximations are known. An approximation due to Welch is given here.
The variable

$$T = \frac{(\bar{X} - \bar{Y}) - (\mu_X - \mu_Y)}{\sqrt{\dfrac{s_X^2}{n_x} + \dfrac{s_Y^2}{n_y}}}$$

is approximately a t variable with ν degrees of freedom, where

$$\nu = \frac{\left(\dfrac{s_X^2}{n_x} + \dfrac{s_Y^2}{n_y}\right)^2}{\dfrac{\left(\dfrac{s_X^2}{n_x}\right)^2}{n_x - 1} + \dfrac{\left(\dfrac{s_Y^2}{n_y}\right)^2}{n_y - 1}}$$

It cannot be emphasized too strongly that the exact probability distribution of T is unknown. ∎

Using the data in Example 10.5, we find $\nu = 17.660$ so we must use a t variable with 17 degrees of freedom. (We must always use the greatest integer less than or equal to ν; otherwise, the sample sizes are artificially increased.) This gives $T_{17} = -3.09$, a result quite comparable to previous results. The Welch approximation will make a very significant difference if the population variances are quite disparate.
It is not difficult to see that , as $n_x \to \infty$ and $n_y \to \infty$,

$$T = \frac{(\bar{X} - \bar{Y}) - (\mu_X - \mu_Y)}{\sqrt{\dfrac{s_X^2}{n_x} + \dfrac{s_Y^2}{n_y}}} \to \frac{(\bar{X} - \bar{Y}) - (\mu_X - \mu_Y)}{\sqrt{\dfrac{\sigma_X^2}{n_x} + \dfrac{\sigma_Y^2}{n_y}}} = z$$

Certainly if each of the sample sizes exceeds 30, the normal approximation will be a very good one. However, it is very dangerous to assume normality for small samples if their population variances are quite different. In that case, the normal approximation is to be avoided. Computer programs such as Minitab make it easy to do the exact calculations involved, regardless of sample size. This is a safe and prudent route to follow in this circumstance.

The tests given here heavily depend on the relationship between the variances as well as the normality of the underlying distributions. If these assumptions are not true, no exact or approximate tests are known for this situation.

Often T is used, but the *minimum* of $n_x - 1$ and $n_y - 1$ is used for the degrees of freedom. If this advice is followed, then in the example we have been using, $T = -3.09$ with 14 degrees of freedom giving a p – value of 0.008. Our approximation in Example 10.5 allows us to use 17 degrees of freedom, making the p–value 0.007. The difference is not great in this case.

CONCLUSIONS

Three new, and very useful, probability distributions have been introduced here that have been used in testing a single variance, a single mean when the population variance is unknown, and two variances. This culminates in several tests on the difference between two means when the population variances are unknown and both when the population variances are assumed to be equal and when they are assumed to be unequal.

We now study a very important application of much of what we have discussed in *statistical process control*.

EXPLORATIONS

1. Two samples of students taking the Scholastic Aptitude Test (SAT) gave the following data:

$$n_1 = 34, \quad \bar{x}_1 = 563, \quad s_1^2 = 80.64$$

$$n_2 = 28, \quad \bar{x}_2 = 602, \quad s_2^2 = 121.68$$

Assume that the samples arise from normal populations.
(a) Test $H_0 : \mu_1 = 540$ against $H_1 : \mu_1 > 540$
(b) Find a 90% confidence interval for σ_2^2.
(c) Test $H_0 : \mu_1 = \mu_2$ assuming
    (i) $\sigma_1^2 = \sigma_2^2$. Is this assumption supported by the data?
    (ii) $\sigma_1^2 \neq \sigma_2^2$. Is this assumption supported by the data?

Chapter 11

Statistical Process Control

CHAPTER OBJECTIVES:

- to introduce control charts
- to show how to estimate the unknown σ
- to show control charts for means, proportion defective, and number of defectives
- to learn about acceptance sampling and how this may improve the quality of manufactured products.

Statistics has become an important subject because, among other contributions to our lives, it has improved the quality of the products we purchase and use. Statistical analysis has become a centerpiece of manufacturing. In this chapter, we want to explore some of the ways in which statistics does this. We first look at *control charts*.

CONTROL CHARTS

EXAMPLE 11.1   *Data from a Production Line*

A manufacturing process is subject to periodic inspection. Samples of five items are selected from the production line periodically. A size measurement is taken and the mean of the five observations is recorded. The means are plotted on a graph as a function of time. The result is shown in Figure 11.1.

The means shown in Figure 11.1 certainly show considerable variation. Observation 15, for example, appears to be greater than any of the other observations while the 19th observation appears to be less than the others. The variability of the means appears to have decreased after observation 20 as well. But, are these apparent changes in the process statistically significant? If so, the process may have undergone some significant changes that may lead to an investigation of the production process. However, processes show random variation arising from a number

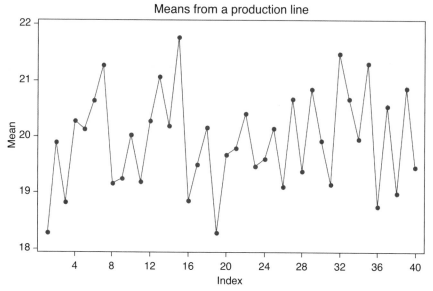

Figure 11.1

of sources, and we may simply be observing this random behavior that occurs in all production processes.

It would help, of course, if we were to know the true mean of the observations μ as well as the standard deviation σ, neither of these quantities is known. How can we proceed in judging the process?

Forty samples, each of size 5, were selected. Some of the samples chosen and summary statistics from these samples are given in Table 11.1. The observations are denoted by $x1$, $x2$, $x3$, $x4$, and $x5$.

Table 11.1

$x1$	$x2$	$x3$	$x4$	$x5$	Mean	Standard Deviation	Range
14.4357	16.3964	22.5443	23.2557	14.7566	18.2777	4.29191	8.81991
21.1211	17.5008	19.6445	19.9313	21.2559	19.8907	1.51258	3.75511
19.9250	17.3215	19.9991	13.4993	23.4052	18.8300	3.68069	9.90589
22.8812	18.8140	21.7421	17.3827	20.5589	20.2758	2.20938	5.49851
19.8692	22.2468	16.5286	20.3042	21.7208	20.1339	2.24052	5.71821
20.5668	20.9625	21.3229	19.1754	21.2247	20.6504	0.87495	2.14750
24.5589	18.9837	23.1692	19.1203	20.5269	21.2718	2.49118	5.57512
23.0998	18.7467	19.7165	17.8197	16.4408	19.1647	2.50961	6.65895
20.8545	18.8997	18.8628	21.7777	15.8875	19.2564	2.26620	5.89023
⋮	⋮	⋮	⋮	⋮	⋮	⋮	⋮
19.4660	19.8596	19.6703	17.3478	20.9661	19.4619	1.31656	3.61823

For each sample, the mean, standard deviation, and range (the difference between the largest observation and the smallest observation) were calculated.

We know the central limit theorem shows that the sample mean $\bar{x}$ follows a normal distribution with mean μ and standard deviation $\sigma/\sqrt{n}$. In this case, n is the sample size 5. It is natural to estimate μ by the mean of the sample means $\bar{\bar{x}}$. In this case, using all 40 sample means, $\bar{\bar{x}} = 19.933$. Now if we could estimate σ, the standard deviation, we might say that a mean greater than $\bar{\bar{x}} + 3(\sigma/\sqrt{n})$ or a mean less than $\bar{\bar{x}} - 3(\sigma/\sqrt{n})$ would be unusual since, for a normal distribution, we know that the probability an observation is outside these limits is 0.0026998, a very unlikely event. While we do not know σ, we might assume, correctly, that the sample standard deviations and the sample ranges may aid us in estimating σ. We now show two ways to estimate σ based on each of these sample statistics. ∎

ESTIMATING σ USING THE SAMPLE STANDARD DEVIATIONS

In Table 11.1, the standard deviation is calculated for each of the samples shown there. The sample standard deviations give us some information about the process standard deviation σ. It is known that the mean of the sample standard deviations $\bar{s}$ can be adjusted to provide an estimate for the standard deviation $\hat{\sigma}$. In fact,

$$\hat{\sigma} = \frac{\bar{s}}{c_4}$$

where the adjustment factor c_4 depends on the sample size. The adjustment c_4 ensures that the expected value of $\hat{\sigma}$ is the unknown σ. Such estimates are called *unbiased*. Table 11.2 shows some values of this quantity.

Table 11.2

n	c_4
2	0.797885
3	0.886227
4	0.921318
5	0.939986
6	0.951533
7	0.959369
8	0.965030
9	0.969311
10	0.972659
11	0.975350
12	0.977559
13	0.979406
14	0.980971
15	0.982316
16	0.983484
17	0.984506

In this case, $\bar{s} = 1.874$ and $c_4 = 0.939986$, so our estimate is

$$\hat{\sigma} = \frac{1.874}{0.939986} = 1.99365$$

This means that the limits we suggested, $\bar{\bar{x}} \pm 3(\sigma/\sqrt{n})$ become $19.933 - 3 \cdot 1.99365/\sqrt{5} = 17.2582$ and $19.933 + 3 \cdot 1.99365/\sqrt{5} = 22.6078$. These are called *upper and lower control limits* (UCL and LCL). If we show these on the graph in Figure 11.1, we produce Figure 11.2.

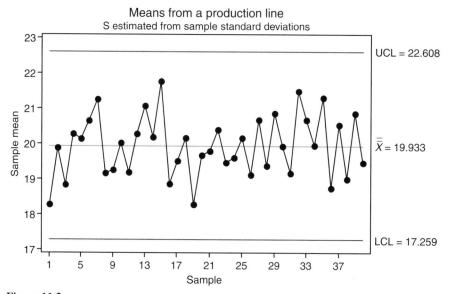

Figure 11.2

None of the observations are outside the control limits.

The use of three standard deviations as control limits is a very conservative choice. The fact that the limits are exceeded so rarely is an argument in its favor since the production line would be shut down or investigated very infrequently. It is possible, of course, to select other control limits. The limits $\bar{x} - 2 \cdot \sigma/\sqrt{n}$ and $\bar{x} + 2 \cdot \sigma/\sqrt{n}$ would be exceeded roughly 4.55% of the time. In this case, these limits are

$$19.933 - 2 \cdot \frac{1.99365}{\sqrt{5}} = 18.1498$$

and

$$19.933 + 2 \cdot \frac{1.99365}{\sqrt{5}} = 21.7161$$

The resulting control chart is shown in Figure 11.3.

Now the 15th observation is a bit beyond the upper control limit.

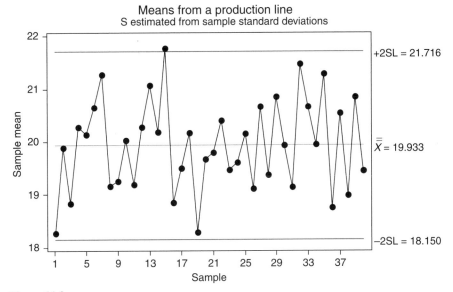

Figure 11.3

ESTIMATING σ USING THE SAMPLE RANGES

The sample ranges can also be used to estimate σ. The mean range $\overline{R}$ must be adjusted to provide an unbiased estimate of σ. It is a fact that an unbiased estimate of σ is

$$\widehat{\sigma} = \frac{\overline{R}}{d_2}$$

where d_2 depends on the sample size. Table 11.3 gives some values of d_2.

Table 11.3

n	d_2
2	1.128
3	1.693
4	2.059
5	2.326
6	2.534
7	2.704
8	2.847
9	2.970
10	3.078

In this case, $\overline{R} = 4.536$ and $d_2 = 2.326$, so our estimate of σ is

$$\hat{\sigma} = \frac{4.536}{2.326} = 1.950\,13$$

and this produces three sigma control limits at $19.933 - 2 \cdot 1.95013/\sqrt{5} = 18.189$ and $19.933 + 2 \cdot 1.95013/\sqrt{5} = 21.678$. The resulting control chart is shown in Figure 11.4.

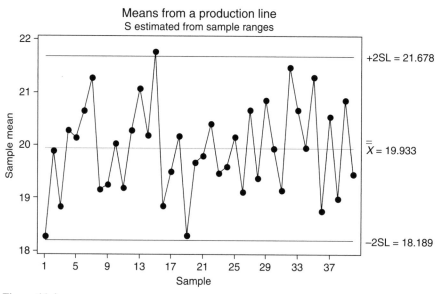

Figure 11.4

The two control charts do not differ much in this case. It is easier on the production floor to calculate $\overline{R}$, but both methods are used with some frequency.

The control charts here were produced using Minitab that allows several methods for estimating σ as well as great flexibility in using various multiples of σ as control limits. It is possible to produce control charts for statistics other than the sample mean, but we will not discuss those charts here. These are examples of control charts for *variables*.

It is natural in our example to use the sample mean as a statistic since we made measurements on each of the samples as they emerged from the production process. It may be, however, that the production items are either usable or defective; in that case, we call the resulting control charts as control charts for *attributes*.

CONTROL CHARTS FOR ATTRIBUTES

EXAMPLE 11.2   *Metal Plating Data*

Thirty samples of size 50 each are selected from a manufacturing process involving plating a metal. The data in Table 11.4 give the number of parts that showed a plating defect.

Table 11.4

Sample number	Defects	Sample number	Defects
1	6	16	5
2	4	17	6
3	7	18	1
4	3	19	6
5	1	20	7
6	3	21	12
7	3	22	7
8	1	23	5
9	5	24	2
10	5	25	3
11	2	26	2
12	11	27	2
13	2	28	5
14	3	29	4
15	2	30	4

We are interested in the number of defects in each sample and how this varies from sample to sample as the samples are taken over a period of time. The control chart involved is usually called an *np chart*. We now show how this is constructed. ■

np Control Chart

Let the random variable X denote the number of parts showing plating defects. The random variable X is a binomial random variable whose probability distribution in general is given by

$$P(X = x) = \binom{n}{x} p^x \cdot (1 - p)^{(n-x)}, \quad x = 0, 1, \ldots, n$$

The random variable X is a binomial random variable because a part either shows a defect or it does not show a defect and we assume, correctly or incorrectly, that the parts are produced independently and with constant probability of a defect p. Since 50 observations were taken in each sample, $n = 50$ and so X takes on integer values from 0 to 50 . We know that the mean value of X is np and the standard deviation is $\sqrt{np(1 - p)}$.

Reasonable control limits then might be from LCL $= \overline{X} - 3\sqrt{np(1-p)}$ to UCL $= \overline{X} + 3\sqrt{np(1-p)}$. We can find from the data that the total number of defective parts is 129, so the mean number of defective parts is $\overline{X} = 129/30 = 4.30$, but we do not know the value of p.

A reasonable estimate for p is the total number of defects divided by the total number of parts sampled or $129/(30)(50) = 0.086$. This gives estimates for the control limits as

$$\mathrm{LCL} = \overline{X} - 3\sqrt{n\widehat{p}(1-\widehat{p})} = 4.30 - 3\sqrt{50(0.086)(1-0.086)}$$
$$= -1.6474$$

and

$$\mathrm{UCL} = \overline{X} + 3\sqrt{n\widehat{p}(1-\widehat{p})} = 4.30 + 3\sqrt{50(0.086)(1-0.086)} = 10.25$$

Since X, the number of parts showing defects, cannot be negative, the lower control limit is taken as 0. The resulting control chart, produced by Minitab, is shown in Figure 11.5.

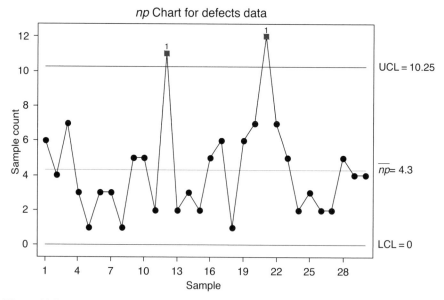

Figure 11.5

It appears that samples 12 and 21 show the process to be out of control. Except for these points, the process is in good control. Figure 11.6 shows the control chart using two sigma limits; none of the points, other than those for samples 12 and 21, are out of control.

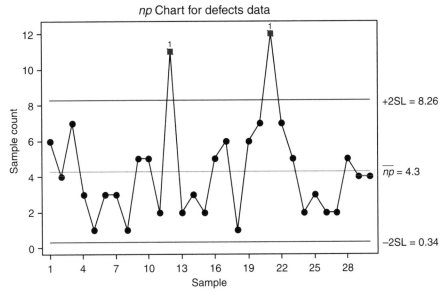

Figure 11.6

p Chart

Due to cost and customer satisfaction, the proportion of the production defective is also of great importance. Denote this random variable as p_s; we know that $p_s = X/n$. Since the random variable X is the number of parts showing defects in our example, X is binomial, and sample size is n, it follows that

$$E(p_s) = E\left(\frac{X}{n}\right) = \frac{E(X)}{n} = \frac{np}{n} = p$$

and

$$\text{Var}(p_s) = \text{Var}\left(\frac{X}{n}\right) = \frac{\text{Var}(X)}{n^2} = \frac{np(1-p)}{n^2} = \frac{p(1-p)}{n}$$

We see that control limits, using three standard deviations, are

$$\text{LCL} = p - 3\sqrt{\frac{p(1-p)}{n}}$$

and

$$\text{UCL} = p + 3\sqrt{\frac{p(1-p)}{n}}$$

but, of course, we do not know the value for p.

A reasonable estimate for p would be the overall proportion defective considering all the samples. This is the estimate used in the previous section, that is, 0.086. This gives control limits as

$$LCL = 0.086 - 3\sqrt{\frac{0.086(1 - 0.086)}{50}} = -0.032948$$

and

$$UCL = 0.086 + 3\sqrt{\frac{0.086(1 - 0.086)}{50}} = 0.2049$$

Zero is used for the lower control limit. The resulting control chart is shown in Figure 11.7.

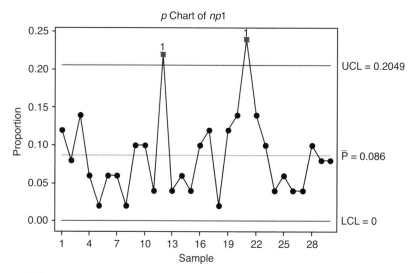

Figure 11.7

This chart gives exactly the same information as the chart shown in Figure 11.6.

SOME CHARACTERISTICS OF CONTROL CHARTS

EXAMPLE 11.3   *Control Limits*

Examine the control chart shown in Figure 11.8. Ignore the numbers 1, 2, and 6 on the chart for the moment, since they will be explained later, but notice that one, two, and three sigma units are displayed.

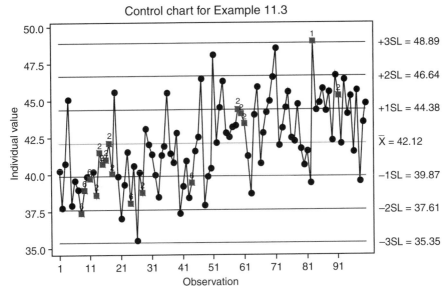

Figure 11.8

This control chart was constructed by selecting 50 samples of size 5 from a normal distribution with mean 40 and standard deviation 5; the next 50 samples of size 5 were chosen from a normal distribution with mean 43.6 and standard deviation 5. The control chart was constructed by taking the mean of the samples. It is apparent from the control chart that something occurred at or around the 50th sample, as indeed it did. But despite the fact that the mean had increased by 3.6 or 72% of the standard deviation, the chart shows only a single point outside three sigma limits, namely, the 83rd, although the 71st and 26th samples are close to the control line. After the 60th sample, the observations become well behaved again, although it is apparent that the mean has increased. The change here was a relatively large one, but using only the three sigma criterion alone, we might not be alarmed. ■

The control charts we have considered offer great insight into production processes since they track the production process in a periodic manner. They have some disadvantages as well; many are slow to discover a change in the production process, especially when the process mean changes by a small proportion of the standard deviation, or when the standard deviation itself changes. In the above example, it is difficult to detect even a large change in the process mean with any certainty. It would be very desirable to detect even small changes in the process mean, and this can be done, but it may take many observations to do so. In the meantime, the process may be essentially out of control without the knowledge of the operators. For this reason, additional tests are performed on the data. We describe some of these now.

SOME ADDITIONAL TESTS FOR CONTROL CHARTS

First, consider the probability of a false reading, that is, an observation falling beyond the three sigma limits entirely due to chance rather than to a real change in the process.

Assuming that the observations are normally distributed, an assumption justified for the sample mean if the sample size is moderate, then the probability of an observation outside the three sigma limits is 0.0027. Such observations are then very rare and when one occurs, we are unlikely to conclude that the observations occurred by chance alone. However, if 50 observations are made, the probability that at least one of them is outside the three sigma limits is $1 - (1 - 0.0027)^{50} = 0.12644$.   This probability increases rapidly, as the data in Table 11.5 indicate. Here n represents the number of observations.

Table 11.5

n	Probability
50	0.126444
100	0.236899
200	0.417677
300	0.555629
400	0.660900
500	0.741233
600	0.802534
700	0.849314
800	0.885011
900	0.912252
1000	0.933039

So an extreme observation becomes almost certain as production continues, although the process in reality has not changed.

Minitab offers eight additional tests on the data; we will describe four of them here, namely, tests 1, 2, 5, and 6. When these tests become significant, Minitab indicates this by putting the appropriate number of the test on the control chart. This explains those numbers on the chart in Figure 11.8.

In general, since the probabilities of these events are quite small, the following situations are to be regarded as cautionary flags for the production process. The calculation of the probabilities involved in most of these tests relies upon the binomial or normal probability distributions. The default values of the constant k in each of these tests can be changed easily in Minitab.

Test 1. Points more than k sigma units from the centerline.

We have used $k = 3$, but Table 11.6 shows probabilities with which a single point is more than k standard deviations from the target mean.

In Figure 11.8, while only one point is beyond the three sigma limit, several are beyond the two sigma limits. These points are indicated by the symbol ■ on the control chart.

Table 11.6

k	Probability
1.0	0.317311
1.5	0.133614
2.0	0.045500
2.5	0.012419
3.0	0.002700

Test 2. k points in a row on the same side of the centerline.

It is common to use $k = 9$. The probability that nine points in a row are on the same side of the centerline is $2(1/2)^9 = 0.00390625$. Table 11.7 shows this probability for $k = 7, 8, \ldots, 11$.

Table 11.7

k	Probability
7	0.01563
8	0.00781
9	0.00391
10	0.00195
11	0.00098

This test fails at samples 13, 14, 16, 17, 18, 28, 60, 61, and 91.

Test 5. At least k out of $k+1$ points in a row more than two sigmas from the centerline.

The quantity k is commonly chosen as 2. Since the probability that one point is more than two sigmas above the centerline is 0.0227501 and since the number of observations outside these limits is a binomial random variable, the probability that at least two out of three observations are more than two sigmas from the centerline and either above or below the centerline is

$$2\sum_{x=2}^{3}\binom{3}{x}(0.0227501)^x(0.9771499)^{3-x} = 0.003035$$

Table 11.8 gives values of this probability for other values of k.

This event occurred at observation 44 in Figure 11.8.

Table 11.8

k	Probability
2	0.003035
3	0.000092
4	0.000003

Test 6. At least k out of $k + 1$ points in a row more than one sigma from the centerline.

The value of k is commonly chosen as 4. Since the probability that one point is more than one sigma above or below the centerline is 0.158655 and since the number of observations outside these limits is a binomial random variable, the probability that at least four out of five observations are more than two sigmas from the centerline is

$$2 \sum_{x=4}^{5} \binom{5}{x} (0.158655)^x (1 - 0.158655)^{5-x} = 0.00553181$$

Table 11.9 gives values of this probability for other values of k.

Table 11.9

k	Probability
2	0.135054
3	0.028147
4	0.005532

This test failed at samples 8, 9, and 11 in our data set.

Computers allow us to calculate the probabilities for each of the tests with relative ease. One could even approximate the probability for the test in question and find a value for k. In test 2, for example, where we seek sequences of points that are on the same side of the centerline, if we wanted a probability of approximately 0.02, Table 11.7 indicates that a run of seven points is sufficient.

The additional tests provide some increased sensitivity for the control chart, but they decrease its simplicity. Simplicity is a desirable feature when production workers monitor the production process.

CONCLUSIONS

We have explored here only a few of the ideas that make statistical analysis and statistical process control, an important part of manufacturing. This topic also shows the power of the computer and especially computer programs dedicated to statistical analysis such as Minitab.

EXPLORATIONS

1. Create a data set similar to that given in Table 11.1. Select samples of size 6 from a $N(25, 3)$ distribution.
   (a) Find the mean, standard deviation, and the range for each sample.
   (b) Use both the sample standard deviations and then the range to estimate σ and show the resulting control charts.

    (c) Carry out the four tests given in the section "Some Additional Tests for Control Chart" and discuss the results.

2. Generate 50 samples of size 40 from a binomial random variable with the probability of a defective item, $p = 0.05$. For each sample, show the number of defective items.

    (a) Construct an np control chart and discuss the results.

    (b) Construct a p chart and discuss the results.

Chapter 12

Nonparametric Methods

CHAPTER OBJECTIVES:

- to learn about hypothesis tests that are not dependent on the parameters of a probability distribution
- to learn about the median and other order statistics
- to use the median in testing hypotheses
- to use runs of successes or failures in hypothesis testing.

INTRODUCTION

The Colorado Rockies national league baseball team early in the 2008 season lost seven games in a row. Is this unusual or would we expect a sequence of seven wins or losses in a row sometime in the regular season of 162 baseball games?

We will answer this question and others related to sequences of successes or failures in this chapter.

We begin with a *nonparametric* test comparing two groups of teachers. In general, *nonparametric* statistical methods refer to statistical methods that do not depend upon the parameters of a distribution, such as the mean and the variance (both of which occur in the definition of the normal distribution, for example), or on the distribution itself.

We begin with two groups of teachers, each one of whom was being considered for an award.

THE RANK SUM TEST

EXAMPLE 12.1   *Two Groups of Teachers*

Twenty-six teachers, from two different schools, were under consideration for a prestigious award (one from each school). Each teacher was ranked by a committee. The results are shown

---

A Probability and Statistics Companion, John J. Kinney
Copyright © 2009 by John Wiley & Sons, Inc.

Table 12.1

Teacher	Score	Rank
A∗	34.5	1
B∗	32.5	2
C∗	29.5	3.5
D	29.5	3.5
E∗	28.6	5
F	27.5	6
G∗	25.0	7
H	24.5	8
I	21.0	9
J	20.5	10
K	19.6	11
L	19.0	12
M	17.0	13.5
N	17.0	13.5
O∗	16.5	15
P∗	15.0	16
Q	14.0	17
R∗	13.0	19
S	13.0	19
T	13.0	19
U	12.5	21
V	9.5	22
W	9.5	23
X	8.5	24
Y	8.0	25
Z	7.0	26

in Table 12.1. Names have been omitted (since this is a real case) and have been replaced by letters. The scores are shown in decreasing order.

The stars (*) after the teacher's names indicate that they came from School I, while those teachers without stars came from School II.

We notice that five of the teachers from School I are clustered near the top of the list.

Could this have occurred by chance or is there a real difference between the teachers at the two schools?

The starred group has 8 teachers, while the unstarred group has 18 teachers. Certainly, the inequality of the sample sizes makes some difference. The usual parametric test, say, on the equality of the means of the two groups is highly dependent upon the samples being selected from normal populations and needs some assumptions concerning the population variances in addition. In this case, these are dubious assumptions to say the least. ∎

So how can we proceed?

Here is one possible procedure called the *rank sum* test.

To carry this test out, we first rank the teachers in order, from 1 to 26. These rankings are shown in the right-hand column of Table 12.1. Note that teachers C and

D each have scores of 29.5, so instead of ranking them as 3 and 4, each is given rank 3.5, the arithmetic mean of the ranks 3 and 4. There are two other instances where the scores are tied, and we have followed the same procedure with those ranks.

Now the ranks of each group are added up; the starred group has a rank sum then of

$$1 + 2 + 3 + 5 + 7 + 15 + 16 + 18.5 = 67.5$$

The sum of all the ranks is

$$1 + 2 + 3 + \cdots + 26 = \frac{26 \cdot 27}{2} = 351$$

so the rank sum for the unstarred teacher group must be $351 - 67.5 = 283.5$.

There is a considerable discrepancy in these rank sums, but, on the contrary, the sample sizes are quite disparate, and so some difference is to be expected. How much of a difference is needed then before we conclude the difference to be statistically significant?

Were we to consider *all the possible* rankings of the group of eight teachers, say, we might then be able to conclude whether or not a sum of 67.5 was statistically significant. It turns out that this is possible. Here is how we can do this.

Suppose then that we examine all the possible rank sums for our eight teachers. These teachers can then be ranked in $\binom{26}{8} = 1,562,275$ ways. It turns out that there are only 145 different rank sums, however. Smaller samples provide interesting classroom examples and can be carried out by hand. In our case, we need a computer to deal with this.

Here is the set of all possible rank sums. Note that the eight teachers could have rank sums ranging from $1 + 2 + 3 + \cdots + 8 = 36$ to $19 + 20 + 21 + \cdots + 26 = 180$.

A table of all the possible values of rank sums is given in Table 12.2.

Table 12.2

{36,37,38,39,40,41,42,43,44,45,46,47,48,49,50,51,52,53,54,55,56,57,58,59,60, 61,62,63,64,65,66,67,68,69,70,71,72,73,74,75,76,77,78,79,80,81,82,83,84,85,86, 87,88,89,90,91,92,93,94,95,96,97,98,99,100,101,102,103,104,105,106,107,108, 109,110,111,112,113,114,115,116,117,118,119,120,121,122,123,124,125,126,127, 128,129,130,131,132,133,134,135,136,137,138,139,140,141,142,143,144,145,146, 147,148,149,150,151,152,153,154,155,156,157,158,159,160,161,162,163,164,165, 166,167,168,169,170,171,172,173,174,175,176,177,178,179,180}

These rank sums, however, do not occur with equal frequency. Table 12.3 shows these frequencies.

Our eight teachers had a rank sum of 67.5. We can then find the rank sums that are at most 67.5. This is $1 + 1 + 2 + 3 + 5 + 7 + \cdots + 2611 = 17244$. So we would conclude that the probability the rank sum is at most 67.5 is $17244/1562275 = 0.01103775$, so it is very rare that a rank sum is less than the rank sum we observed.

Table 12.3

{1,1,2,3,5,7,11,15,22,29,40,52,70,89,116,146,186,230,288,351,432,521,631,752,
900,1060,1252,1461,1707,1972,2281,2611,2991,3395,3853,4338,4883,5453,6087,
6748,7474,8224,9042,9879,10783,11702,12683,13672,14721,15765,16862,17946,
19072,20171,21304,22394,23507,24563,25627,26620,27611,28512,29399,30186,30945,
31590,32200,32684,33125,33434,33692,33814,33885,33814,33692,33434,33125,32684,
32200,31590,30945,30186,29399,28512,27611,26620,25627,24563,23507,22394,21304,
20171,19072,17946,16862,15765,14721,13672,12683,11702,10783,9879,9042,8224,
7474,6748,6087,5453,4883,4338,3853,3395,2991,2611,2281,1972,1707,1461,1252,
1060,900,752,631,521,432,351,288,230,186,146,116,89,70,52,40,29,22,15,11,7,5,
3,2,1,1}

It is interesting to see a graph of all the possible rank sums, as shown in Figure 12.1.

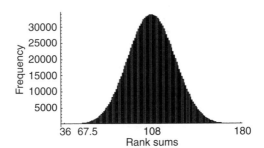

Figure 12.1

Our sum of 67.5 is then in the far left-hand end of this obviously normal curve showing a perhaps surprising occurrence of the normal curve. It arises in many places in probability where one would not usually expect it! Teachers A and D won the awards.

We now turn to some other nonparametric tests and procedures.

ORDER STATISTICS

EXAMPLE 12.2   *Defects in Manufactured Products*

The number of defects in several samples of a manufactured product was observed to be

$$1, 6, 5, 5, 4, 3, 2, 2, 4, 6, 7$$

The mean number of defects is then

$$\frac{1+6+5+5+4+3+2+2+4+6+7}{11} = 4.09$$

■

This mean value is quite sensitive to each of the sample values. For example, if the last observation had been 16 rather than 7, the mean value becomes 4.91, almost a whole unit larger due to this single observation.

So we seek a measure of central tendency in the data that is not so sensitive.

Median

Consider then arranging the data in order of magnitude so that the data become

$$1, 2, 2, 3, 4, 4, 5, 5, 6, 6, 7$$

The *median* is the value in the middle when the data are arranged in order (or the mean of the two middlemost values if the number of observations is even) . Here that value is 4. Now if the final observation becomes 17, for example (or any other value larger than 6), the median remains at 4.

The median is an example of an *order statistic*—values of the data that are determined by the data when the data are arranged in order of magnitude. The smallest value, the *minimum*, is an order statistic as is the largest value, or the *maximum*.

While there is no central limit theorem as there is for the mean, we explore now the probability distribution of the median.

EXAMPLE 12.3   *Samples and the Median*

Samples of size 3 are selected, without replacement, from the set 1,2,3,4,5,6,7.

There are then $\binom{7}{3} = 35$ possible samples. These are shown in Table 12.4 where the median has been calculated for each sample. The samples are arranged in order. Were the samples not

Table 12.4

Sample	Median	Sample	Median
1,2,3	2	2,3,6	3
1,2,4	2	2,3,7	3
1,2,5	2	2,4,5	4
1,2,6	2	2,4,6	4
1,2,7	2	2,4,7	4
1,3,4	3	2,5,6	5
1,3,5	3	2,5,7	5
1,3,6	3	2,6,7	6
1,3,7	3	3,4,5	4
1,4,5	4	3,4,6	4
1,4,6	4	3,4,7	4
1,4,7	4	3,5,6	5
1,5,6	5	3,5,7	5
1,5,7	5	3,6,7	6
1,6,7	6	4,5,6	5
2,3,4	3	4,5,7	5
2,3,5	3	4,6,7	6
		5,6,7	6

arranged in order, each sample would produce $3! = 6$ samples, each with the same median, so we will consider only the ordered samples.

This produces the probability distribution function for the median as shown in Table 12.5.

Table 12.5

Median	Probability
2	5/35
3	8/35
4	9/35
5	8/35
6	5/35

■

The probabilities are easy to find. If the median is k, then one observation must be chosen from the $k - 1$ integers less than k and the remaining observation must be selected from the $7 - k$ integers greater than k, so the probability the median is k then becomes

$$P(\text{median} = k) = \frac{\binom{k-1}{1} \cdot \binom{7-k}{1}}{\binom{7}{3}}, \quad k = 2, 3, 4, 5, 6$$

The expected value of the median is then

$$2 \cdot 5/35 + 3 \cdot 8/35 + 4 \cdot 9/35 + 5 \cdot 8/35 + 6 \cdot 5/35 = 4$$

A graph of this probability distribution is shown in Figure 12.2.

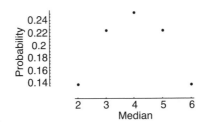

Figure 12.2

This does not show much pattern. In fact, the points here are points on the parabola

$$y = (x - 1)(7 - x)$$

We need a much larger population and sample to show some characteristics of the distribution of the median, and we need a computer for this. The only way for us to show the distribution of the median is to select larger samples than in the above example and to do so from larger populations.

Figure 12.3 shows the distribution of the median when samples of size 5 are selected from $1, 2, 3, \ldots, 100$.

The points on the graph in Figure 12.3 are actually points on the fourth-degree polynomial $y = (x - 1)(x - 2)(100 - x)(99 - x)$.

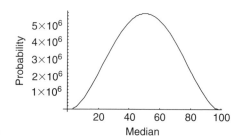

Figure 12.3

Finally, in Figure 12.4 we show the median in samples of size 7 from the population $1, 2, 3, \ldots, 100$.

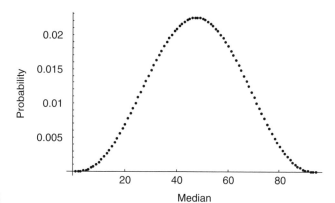

Figure 12.4

There is now little doubt that the median becomes normally distributed. We will not pursue the theory of this but instead consider another order statistic, the maximum.

Maximum

The maximum value in a data set as well as the minimum, the smallest value in the data set, is of great value in statistical quality control where the *range*, the difference between the maximum and the minimum, is easily computed on the production floor from a sample. The range was used in Chapter 11 on statistical process control.

EXAMPLE 12.4   *An Automobile Race*

We return now to two examples that we introduced in Chapter 4.

A latecomer to an automobile race observes cars numbered 6,17, and 45. Assuming the cars to be numbered from 1 to n, what is n, that is, how many cars are in the race?

The question may appear to be a trivial one, but it is not. In World War II, the Germans numbered their tanks in the field. When some tanks were captured and their numbers noted, we were able to estimate the total number of tanks they had.

Clearly, we should use the sample in some way in *estimating* the value for n.

We might use the median that we have been discussing or the mean for the samples. If we refer back to Example 12.3, we saw that the expected value of the median is 4, not a very accurate estimate of the maximum, 7.

Table 12.6 below shows the probability distribution for the mean of the samples in Example 12.3.

Table 12.6

Mean	Frequency
2	1
7/3	1
8/3	2
3	3
10/3	4
11/3	4
4	5
13/3	4
14/3	4
5	4
16/3	3
17/3	1
6	1

So the expected value of the mean is

$$(1/35)[2 \cdot 1 + (7/3) \cdot 1 + (8/3) \cdot 2 + 3 \cdot 3 + (10/3) \cdot 4 + (11/3) \cdot 4 + 4 \cdot 5 + (13/3) \cdot 4$$
$$+ (14/3) \cdot 4 + 5 \cdot 3 + (16/3) \cdot 2 + (17/3) \cdot 1 + 6 \cdot 1] = 4 \qquad \blacksquare$$

This of course was to be expected from our previous experience with the sample mean and the central limit theorem.

Both the median and the mean could be expected, without surprise, to be estimators of the population mean, or a measure of central tendency rather than estimators of the *maximum* of the population, which is what we seek.

It is intuitively clear that if the race cars we saw were numbered 6, 17 and 45, then there are at least 45 cars in the race. But how many more?

Let us look at the probability distribution of the maximum of each sample in Example 12.3. If we do that, we find the probability distribution shown in Table 12.7.

Table 12.7

Maximum	Frequency
3	1
4	3
5	6
6	10
7	15

We find the expected value of the maximum to be

$$\frac{1}{35}[3 \cdot 1 + 4 \cdot 3 + 5 \cdot 6 + 6 \cdot 10 + 7 \cdot 15] = 6$$

This is better than the previous estimators, but can not be the best we can do. For one thing, the maximum only achieves the largest value in the population, 7, with probability 15/35, so with probability 20/35, the maximum of the samples is less than that for the population and yet the maximum of the sample must carry more information with it than the other two values.

We are left with the question, "How should the maximum of the sample be used in estimating the maximum of the population?"

In Figure 12.5, the numbers observed on the cars are shown in a number line.

Figure 12.5   1 6   17   45 n

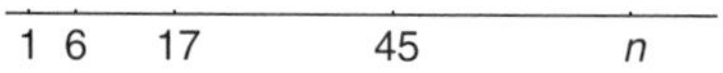

Now we observe that while we saw three cars, we know that cars $1, 2, \ldots, 5$ (a total of 5 cars) must be on the track as are cars numbered $7, 8, \ldots 16$ (a total of 10 cars), as well as cars numbered $18, 19, \ldots, 44$ (a total of 27 cars), so there are $5 + 10 + 27 = 42$ cars that we did not observe.

The three cars we did observe then divide the number line into three parts, which we will call *gaps*. The average size of the gaps is then $42/3 = 14$. Since this is the average gap, it would appear sensible that this is also the gap between the sample maximum, 45, and the unknown population maximum, n. Adding the average gap to the sample maximum gives $45 + 14 = 59$ as our estimate for n.

Before investigating the properties of this estimator, let us note a simple fact. We made heavy use of the numbers 6 and 17 above, but these sizes of these values turn out to be irrelevant. Now suppose that those cars were numbered 14 and 36. The average gap would then be $(13 + 21 + 8)/3 = 42/3 = 14$, the same average gap we found before. In fact, the average gap will be 14 regardless of the size of the two numbers as long as they are less than 45. The reason for this is quite simple: since we observed car number 45, and since we observed 3 cars in total, there must be 42 cars we did not see, giving the average gap as 14. One can also see this by moving the numbers for the two numbers less than 45 back and forth along the number line in Figure 12.5.

So our estimate for n is $45 + (45 - 3)/3 = 59$.

Now to generalize this, keeping our sample size at 3 and supposing the largest observation in the sample is m, our estimate then becomes

$$m + \frac{m - 3}{3} = \frac{4m - 3}{3}$$

Let us see how this works for the samples in Example 12.3. Table 12.8 shows the maximum of each sample and our gap estimator.

Table 12.8

Maximum	Gap estimator	Frequency
3	3	1
4	13/3	3
5	17/3	6
6	7	10
7	25/3	15

This gives the expected value of the gap estimator as

$$(1/35) \cdot (3 \cdot 1 + (13/3) \cdot 3 + (17/3) \cdot 6 + 7 \cdot 10 + (25/3) \cdot 15) = 7$$

the result we desired.

In fact, this estimator can be shown to be *unbiased* (that is, its expected value is the value to be estimated) regardless of the sample size. Suppose our sample is of size k and is selected from the population $1, 2, \ldots, M$ and that the sample maximum is m. Then it can be shown, but not easily, that the expected value for m is not quite M. In fact,

$$E[m] = \sum_{m=k}^{M} m \cdot \frac{\binom{m-1}{k-1}}{\binom{M}{m}} = \frac{k(M+1)}{k+1}$$

So,

$$E\left[\frac{(k+1)m - k}{k} \right] = M$$

We still have a slight problem. In our example, for some samples, the gap method asked us to estimate n as 17/3, obviously not an integer. In that case, we would probably select the integer nearest to 17/3 or 6.

If we do this, our estimator is now distributed as shown in Table 12.9.

We have denoted by square brackets the greatest integer function. The expected value of this estimator is then

$$(1/35) \cdot (3 \cdot 1 + 4 \cdot 3 + 6 \cdot 6 + 7 \cdot 10 + 8 \cdot 15) = 6.89$$

Table 12.9

Maximum	[Gap estimation]	Frequency
3	3	1
4	4	3
5	6	6
6	7	10
7	8	15

Now if we use the greatest integer function here, we estimate M as 7 again. This estimator appears to be close to unbiased, but that is not at all easy to show.

This method was actually used in World War II and proved to be surprisingly accurate in solving the German Tank problem.

Runs

Finally, in this chapter, we consider runs of luck (or runs of ill-fortune).

EXAMPLE 12.5   *Losing Streaks in Baseball*

The Colorado Rockies, a National League baseball team, lost seven games in a row early in the 2008 season. Is this unusual or can a sequence of seven losses (or wins) be expected during a regular season of 162 games?

A *run* is a series of like events in a sequence of events. For example, a computer program (this can also be done with a spreadsheet) gave the following sequence of 20 1's and 0's:

$$\{0, 0, 1, 1, 1, 1, 1, 1, 0, 0, 0, 1, 1, 0, 0, 1, 0, 0, 1, 0\}$$

The sequence starts with a run of two zeros, followed by six ones, three zeros, and so on.

There are nine runs in all and their lengths are given in Table 12.10

The expected number of runs is then

$$(1/9)(1 \cdot 3 + 2 \cdot 4 + 3 \cdot 1 + 4 \cdot 0 + 5 \cdot 0 + 6 \cdot 1) = 2.22$$

Table 12.10

Length	Frequency
1	3
2	4
3	1
4	0
5	0
6	1

It is not at all clear from this small example that this is a typical situation at all. So we simulated 1000 sequences of 20 ones and zeros and counted the total number of runs and their lengths.

To write a computer program to do this, it is easiest to count the total number of runs first. To do this, scan the sequence from left to right, comparing adjacent entries. When they differ, add one to the number of runs until the sequence is completely scanned. For example, in the sequence

$$\{0, 0, 1, 1, 1, 1, 1, 1, 0, 0, 0, 1, 1, 0, 0, 1, 0, 0, 1, 0\}$$

we find the first time adjacent entries are not equal occurs at the third entry, then at the ninth entry, then at the thirteenth entry, and so on. This gives one less than the total number of entries since the first run of two zeros is not counted.

To find the lengths of the individual runs, start with a vector of ones of length the total number of runs. In this example, we begin with the vector

$$\{1, 1, 1, 1, 1, 1, 1, 1, 1\}$$

Scan the sequence again, adding one when adjacent entries are alike and skipping to the next one in the ones vector when adjacent entries differ, and then adding one until the entries differ and continuing though the vector of ones.

In our example, this will produce the vector $\{2, 6, 3, 2, 2, 1, 2, 1, 1\}$. (The sum of the entries must be 20.)

Figure 12.6 shows a bar chart of the number of runs from the 1000 samples.

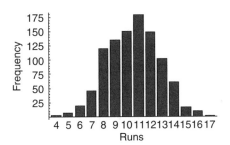

Figure 12.6

Now we return to our baseball example where the Colorado Rockies lost seven games in a row in the 2008 regular baseball season.

We simulated 300 seasons, each with 162 games. Here is a typical simulated year where 0 represents a loss and 1 represents a win.

$$\{0, 1, 1, 1, 0, 0, 1, 0, 1, 1, 0, 0, 1, 0, 0, 0, 1, 1, 0, 1, 0, 1, 0, 0, 1,$$
$$1, 0, 1, 1, 1, 1, 0, 1, 1, 1, 0, 1, 0, 0, 0, 0, 0, 0, 0, 0, 1, 1, 1, 1, 1, 1, 0,$$
$$0, 0, 0, 0, 1, 1, 0, 1, 1, 0, 1, 0, 1, 0, 0, 1, 1, 1, 0, 0, 0, 0, 0, 1, 0, 0,$$
$$1, 0, 1, 0, 0, 1, 0, 0, 0, 1, 1, 0, 0, 0, 1, 0, 1, 0, 0, 1, 0, 0, 0, 0, 0, 1,$$
$$1, 0, 0, 0, 0, 1, 1, 1, 1, 1, 1, 0, 0, 0, 0, 0, 1, 0, 0, 1, 1, 1, 1, 0, 1, 0, 1,$$
$$1, 0, 1, 0, 1, 0, 0, 1, 0, 1, 1, 1, 1, 0, 0, 1, 1, 1, 1, 0, 1, 1, 0, 0, 1, 1,$$
$$1, 0, 1, 0, 0, 1, 1\}$$

We then counted the number of runs in the 300 samples. Table 12.11 shows the number of runs and Figure 12.7 shows a graph of these results.

Table 12.11

{74, 88, 89, 85, 79, 92, 81, 82, 78, 83, 86, 86, 83, 73, 78, 73, 86, 75, 83,
79, 87, 81, 86, 91, 81, 74, 80, 83, 82, 83, 76, 84, 76, 84, 79, 78, 71, 72,
86, 82, 87, 87, 90, 91, 80, 75, 87, 72, 73, 84, 84, 91, 81, 83, 73, 82, 86,
80, 82, 88, 86, 84, 84, 76, 81, 85, 79, 69, 72, 85, 81, 80, 82, 85, 88, 77,
86, 68, 84, 76, 82, 75, 80, 85, 80, 84, 84, 76, 87, 89, 94, 84, 84, 81, 67,
92, 78, 75, 78, 68, 92, 73, 75, 79, 79, 83, 84, 77, 83, 85, 80, 75, 82, 83,
84, 76, 93, 94, 72, 73, 90, 89, 91, 83, 99, 76, 91, 87, 93, 88, 86, 96, 74,
78, 83, 75, 80, 80, 78, 73, 88, 76, 76, 85, 84, 91, 75, 74, 89, 82, 86, 83,
87, 85, 84, 86, 72, 83, 80, 81, 74, 82, 85, 85, 81, 71, 78, 81, 72, 85, 80,
87, 83, 78, 75, 79, 83, 79, 76, 79, 76, 94, 85, 80, 87, 70, 81, 82, 84, 90,
92, 85, 66, 73, 81, 74, 87, 90, 77, 77}

The mean number of runs is 81.595 with standard deviation 6.1587.

The simulation showed $135 + 49 + 26 + 15 + 55 + 5 + 3 + 2 + 1 = 291$ samples having winning or losing streaks of 7 games or more out of a total of 16, 319 runs or a probability of 0.017832.

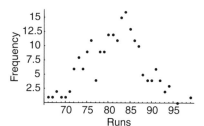

Figure 12.7

Table 12.12 shows the frequency with which runs of a given length occurred, and Figure 12.8 is a graph of these frequencies.

This example shows the power of the computer in investigating the theory of runs. The mathematical theory is beyond our scope here, but there are some interesting theoretical results we can find using some simple combinatorial ideas. ■

Some Theory of Runs

Consider for a moment, the arrangement below of *'s contained in cells limited by bars (|'s):

$$|| ** || *** || * | **** |$$

There are eight cells, only four of which are occupied. There are nine bars in total, but we need one at each end of the sequence to define the first and last cell. Consider the bars to be all alike and the stars to be all alike.

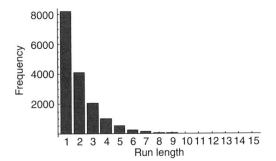

Figure 12.8

Table 12.12

Run length	Frequency
1	8201
2	4086
3	2055
4	983
5	506
6	247
7	135
8	49
9	26
10	15
11	5
12	5
13	3
14	2
15	1

How many different arrangements are there? We can not alter the bars at the ends, so we have 7 bars and 10 stars to arrange. This can be done in $\frac{(7+10)!}{7!10!} = 19,448$ different ways.

At present, the cells will become places into which we put symbols for runs. But, of course, we can not have any empty cells, so we pursue that situation.

We have 10 stars in our example and 8 cells. If we wish to have each of the cells occupied, we could, for example, have this arrangement.

$$| * | ** | * | ** | * | * | * | * |$$

There are other possibilities now. How many such arrangements are there?

First, place a star in each cell. This leaves us with 2 stars and 7 bars which can be arranged in $\frac{(7+2)!}{7!2!} = 36$ different ways.

Now we generalize the situation. Suppose there are n cells and r stars. The n cells are defined by $n + 1$ bars, but two of these are fixed at the ends, so we have $n - 1$ bars that can be arranged along with the r stars. This can be done in

$$\frac{(n - 1 + r)!}{(n - 1)! r!} = \binom{n - 1 + r}{n - 1}$$

different ways.

Now if each cell is to contain at least one star, place one star in each cell. This leaves us with $r - n$ stars to put into n cells. This can then be done in

$$\binom{n - 1 + r - n}{n - 1} = \binom{r - 1}{n - 1}$$

ways.

This formula can then be used to count the number of runs from two sets of objects, say x's and y's. We must distinguish, however, between an even number of runs and an odd number of runs.

Suppose that there are n_x x's and n_y y's and that we have an even number of runs, so say there are $2k$ runs. Then there must be k runs of the x's and k runs of the y's.

There are $\binom{n_x + n_y}{n_x}$ different arrangements of the x's and y's. Let us fill the k cells with x's first, leaving no cell empty. This can be done in $\binom{n_x - 1}{k - 1}$ ways. Then we must fill the k cells for the y's so that no cell is empty. This is done in $\binom{n_y - 1}{k - 1}$ ways. So we can fill all the cells in $\binom{n_x - 1}{k - 1} \binom{n_y - 1}{k - 1}$ ways. This is also the number of ways we can fill all the cells if were we to choose the y's first.

So if R is the random variable denoting the number of runs, then

$$P(R = 2k) = \frac{2 \binom{n_x - 1}{k - 1} \binom{n_y - 1}{k - 1}}{\binom{n_x + n_y}{n_x}}$$

Now consider the case where the number of runs is odd, so $R = 2k + 1$. This means that the number of runs of one of the letters is k and the number of runs of the other letter is $k + 1$. It follows that

$$P(R = 2k + 1) = \frac{\binom{n_x - 1}{k - 1} \binom{n_y - 1}{k} + \binom{n_x - 1}{k} \binom{n_y - 1}{k - 1}}{\binom{n_x + n_y}{n_x}}$$

We will not show it, but it can be shown that

$$E(R) = \frac{2 n_x n_y}{n_x + n_y} + 1$$

$$Var(R) = \frac{2 n_x n_y (2 n_x n_y - n_x - n_y)}{(n_x + n_y)^2 (n_x + n_y - 1)}$$

In the baseline example, assuming that $n_x = n_y = 81$, the formulas give $E(R) = 82$ and

$$Var(R) = \frac{2 \cdot 81 \cdot 81 \cdot (2 \cdot 81 \cdot 81 - 81 - 81)}{(81 + 81)^2(81 + 81 - 1)} = 40.248$$

so the standard deviation is $\sqrt{40.248} = 6.344$, results that are very close to those found in our simulation.

EXAMPLE 12.6

Suppose $n_x = 5$ and $n_y = 8$. This is a fairly large example to do by hand since there are $\binom{5+8}{5} = 1287$ different orders in which the letters could appear.

Table 12.13 shows all the probabilities of the numbers of runs multiplied by 1287, and Figure 12.9 shows a graph of the resulting probabilities. ■

Table 12.13

Runs	Probability
2	2
3	11
4	56
5	126
6	294
7	280
8	175
9	70
10	21

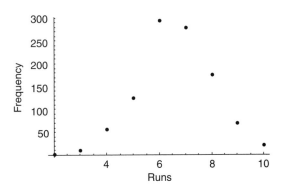

Figure 12.9

EXAMPLE 12.7   *Even and Odd Runs*

And now we return to our baseball example. It is possible to separate the baseball seasons for which the number of runs was even from those for which it was odd. Figures 12.10 and 12.11 show the respective distributions.

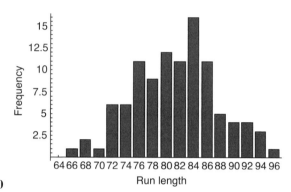

Figure 12.10

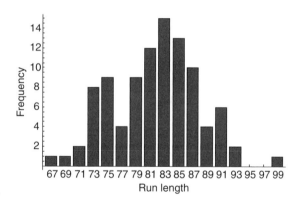

Figure 12.11

It is interesting to compare some of the statistics for the years with an even number of runs with those having an odd number of runs. There are 149 years with an even number of runs with mean 80.9799 and standard deviation 6.0966. There are 151 years with an odd number of runs with mean 81.5033 and standard deviation 6.4837.   ∎

CONCLUSIONS

In this chapter, we have introduced some nonparametric statistical methods; that is, methods that do not make distributional assumptions. We have used the median and the maximum in particular and have used them in statistical testing; there are many other nonparametric tests and the reader is referred to many texts in this area. We pursued the theory of runs and applied the results to winning and losing streaks in baseball.

EXPLORATIONS

1. Find all the samples of size 4 chosen without replacement from the set 1, 2, 3, ..., 8. Find the probability distributions of
   (a) the sample mean,
   (b) the sample median,
   (c) the sample maximum.

2. Choose all the possible samples of size 3 selected without replacement from the set 1, 2, 3, ..., 10. Use the "average gap" method for each sample to estimate the maximum of the population and show its probability distribution.

3. The text indicates a procedure that can be used with a computer to count the number of runs in a sample. Produce 200 samples of size 10 (using the symbols 0 and 1) and show the probability distribution of the number of runs.

4. Show all the permutations of five integers. Suppose the integers 1, 2, and 3 are from population I and the integers 4 and 5 are from population II. Find the probability distribution of all the possible rank sums.

Chapter 13

Least Squares, Medians, and the Indy 500

CHAPTER OBJECTIVES:

- to show two procedures for approximating bivariate data with straight lines one of which uses medians
- to find some surprising connections between geometry and data analysis
- to find the least squares regression line without calculus
- to see an interesting use of an elliptic paraboloid
- to show how the equations of straight lines and their intersections can be used in a practical situation
- to use the properties of median lines in triangles that can be used in data analysis.

INTRODUCTION

We often summarize a set of data by a single number such as the mean, median, range, standard deviation, and many other measures. We now turn our attention to the analysis of a data set with two variables by an equation. We ask, "Can a bivariate data set be described by an equation?" As an example, consider the following data set that represents a very small study of blood pressure and age.

Age	Blood pressure
35	114
45	124
55	143
65	158
75	166

A Probability and Statistics Companion, John J. Kinney
Copyright © 2009 by John Wiley & Sons, Inc.

We seek here an equation that approximates and summarizes the data. First, let us look at a graph of the data. This is shown in Figure 13.1.

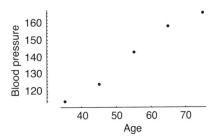

Figure 13.1

It would appear that the data could be well approximated by a straight line.

We might guess some straight lines that might fit the data well. For example, we might try the straight line $y = 60 + 1.2x$, where y is blood pressure and x is age. How well does this line fit the data? Let us consider the predictions this line makes, call them $\hat{y}_i$, and the observed values, say y_i. We have shown these values and the discrepancies, $y - \hat{y}$, in Table 13.1.

Table 13.1

x	y	$\hat{y}$	$y - \hat{y}$
35	114	102	12
45	124	114	10
55	143	126	17
65	158	138	20
75	166	150	16

The discrepancies, or what are commonly called *errors* or *residuals*, happen to be all positive in this case, but that is not always so. So how are we to measure the adequacy of this straight line approximation or fit? Sometimes, of course, the positive residuals will offset the negative residuals, so adding up the residuals can be quite misleading. To avoid this complication, it is customary to square the residuals before adding them up. If we do that in this case we get 1189, but we do not know if that can be improved upon. So let us try some other combinations of straight lines.

First, suppose the line is of the form $y_i = \alpha + \beta x_i$. Although the details of the calculations have not been shown, Table 13.2 shows some values for the sum of squares, $SS = \sum_{i=1}^{5}(y - \hat{y})^2$, and various choices for α and β.

One could continue in this way, trying various combinations of α and β until a minimum is reached. The minimum in this case, as we shall soon see, occurs when $\alpha = 65.1$ and $\beta = 1.38$, producing a minimum sum of squares of 65.1. But trial and error is a very inefficient way to determine the minimum sum of squares and is feasible in this case because the data set consists of only five data points. It is clearly nearly impossible, even with a computer, when we consider subsequently an

Table 13.2

α	β	SS
55	1	4981
60	1.2	1189
65	1.3	139.25
70	1.4	212
75	1.4	637

example consisting of all the winning speeds at the Indianapolis 500-mile race, a data set consisting of 91 data points.

In our small case, it is possible to examine a surface showing the sum of squares (SS) as a function of α and β. A graph of SS $= \sum_{i=1}^{n}(y_i - \alpha - \beta x_i)^2$ is shown in Figure 13.2. It can be shown that the surface is an elliptic paraboloid, that is, the intersections of the surface with vertical planes are parabolas and the intersections of the surface with horizontal planes are ellipses.

It is clear that SS does reach a minimum, although it is graphically difficult to determine the exact values of α and β that produce that minimum. We now show an

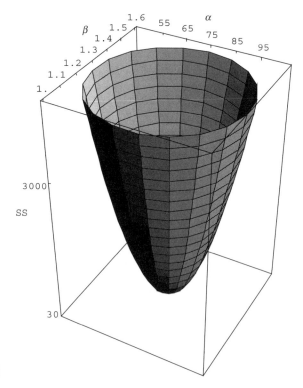

Figure 13.2

algebraic method for determining the values of α and β that minimize the sum of squares.

LEAST SQUARES

Minimizing the sum of squares of the deviations of the predicted values from the observed values is known as the principle of *least squares.*

We now show how to determine the values of α and β that minimize the sum of squares.

Principle of Least Squares

Estimate α and β by those values that minimize $SS = \sum_{i=1}^{n}(y_i - \alpha - \beta x_i)^2$.

Suppose that we have a set of data $\{x_i, y_i\}$ where $i = 1, \cdots, n$. Our straight line is $y_i = \alpha + \beta x_i$ and our sum of squares is $SS = \sum_{i=1}^{n}(y_i - \alpha - \beta x_i)^2$. The values that minimize this sum of squares are denoted by $\hat{\alpha}$ and $\hat{\beta}$.

The least squares line then estimates the value of y for a particular value of x as $\hat{y}_i = \hat{\alpha} + \hat{\beta} x_i$. So the principle of least squares says that we estimate the intercept (α) and the slope (β) of the line by those values that minimize the sum of squares of the *residuals*, $y_i - \alpha - \beta x_i$; the differences between the observed values, the y_i; and the values predicted by the line, $\alpha + \beta x_i$.

We now look at the situation in general. We begin with

$$SS = \sum_{i=1}^{n}(y_i - \alpha - \beta x_i)^2$$

which can be written as

$$SS = \sum_{i=1}^{n} y_i^2 + n\alpha^2 + \beta^2 \sum_{i=1}^{n} x_i^2 - 2\alpha \sum_{i=1}^{n} y_i - 2\beta \sum_{i=1}^{n} x_i y_i + 2\alpha\beta \sum_{i=1}^{n} x_i$$

From Figure 13.2 and the above equation, we find that if we hold β fixed, the intersection is a parabola that has a minimum value.

So we hold β fixed and write SS as a function of α alone:

$$SS = n\alpha^2 - 2\alpha \left(\sum_{i=1}^{n} y_i - \beta \sum_{i=1}^{n} x_i \right) + \sum_{i=1}^{n} y_i^2 + \beta^2 \sum_{i=1}^{n} x_i^2 - 2\beta \sum_{i=1}^{n} x_i y_i$$

Now factor out the factor of n and noting that $\sum_{i=1}^{n} y_i/n = \bar{y}$, with a similar result for $\bar{x}$, and adding and subtracting $n(\bar{y} - \beta\bar{x})^2$ (thus completing the square), we can write

$$SS = n[\alpha^2 - 2\alpha(\bar{y} - \beta\bar{x}) + (\bar{y} - \beta\bar{x})^2] - n(\bar{y} - \beta\bar{x})^2 + \sum_{i=1}^{n} y_i^2$$

$$+\beta^2 \sum_{i=1}^{n} x_i^2 - 2\beta \sum_{i=1}^{n} x_i y_i$$

or

$$SS = n[\alpha - (\bar{y} - \beta\bar{x})]^2 - n(\bar{y} - \beta\bar{x})^2 + \sum_{i=1}^{n} y_i^2 + \beta^2 \sum_{i=1}^{n} x_i^2 - 2\beta \sum_{i=1}^{n} x_i y_i$$

Now since β is fixed and n is positive, the minimum value for SS occurs when $\hat{\alpha} = \bar{y} - \beta\bar{x}$. Note that in our example, $\bar{y} = 141$ and $\bar{x} = 55$, giving the estimate for $\hat{\alpha}$ as $141 - 55\hat{\beta}$. However, we now have a general form for our estimate of α.

Now we find an estimate for β. Hold α fixed. We can write, using the above result for $\hat{\alpha}$,

$$SS = \sum_{i=1}^{n}(y_i - (\bar{y} - \beta\bar{x}) - \beta x_i)^2 = \sum_{i=1}^{n}[(y_i - \bar{y}) - \beta(x_i - \bar{x})]^2$$

$$= \beta^2 \sum_{i=1}^{n}(x_i - \bar{x})^2 - 2\beta \sum_{i=1}^{n}(x_i - \bar{x})(y_i - \bar{y}) + \sum_{i=1}^{n}(y_i - \bar{y})^2$$

Now factor out $\sum_{i=1}^{n}(x_i - \bar{x})^2$ and complete the square to find

$$SS = \sum_{i=1}^{n}(x_i - \bar{x})^2 \left[\beta^2 - 2\frac{\sum_{i=1}^{n}(x_i - \bar{x})(y_i - \bar{y})}{\sum_{i=1}^{n}(x_i - \bar{x})^2} + \left(\frac{\sum_{i=1}^{n}(x_i - \bar{x})(y_i - \bar{y})}{\sum_{i=1}^{n}(x_i - \bar{x})^2} \right)^2 \right]$$

$$- \sum_{i=1}^{n}(x_i - \bar{x})^2 \left(\frac{\sum_{i=1}^{n}(x_i - \bar{x})(y_i - \bar{y})}{\sum_{i=1}^{n}(x_i - \bar{x})^2} \right)^2 + \sum_{i=1}^{n}(y_i - \bar{y})^2$$

which can be written as

$$SS = \sum_{i=1}^{n}(x_i - \bar{x})^2 \left[\beta - \frac{\sum_{i=1}^{n}(x_i - \bar{x})(y_i - \bar{y})}{\sum_{i=1}^{n}(x_i - \bar{x})^2}\right]^2$$

$$- \sum_{i=1}^{n}(x_i - \bar{x})^2 \left(\frac{\sum_{i=1}^{n}(x_i - \bar{x})(y_i - \bar{y})}{\sum_{i=1}^{n}(x_i - \bar{x})^2}\right)^2 + \sum_{i=1}^{n}(y_i - \bar{y})^2$$

showing that the minimum value of SS is attained when

$$\hat{\beta} = \frac{\sum_{i=1}^{n}(x_i - \bar{x})(y_i - \bar{y})}{\sum_{i=1}^{n}(x_i - \bar{x})^2}$$

We have found that the minimum value of SS is achieved when

$$\hat{\beta} = \frac{\sum_{i=1}^{n}(x_i - \bar{x})(y_i - \bar{y})}{\sum_{i=1}^{n}(x_i - \bar{x})^2}$$

and

$$\hat{\alpha} = \bar{y} - \hat{\beta}\bar{x}$$

We have written these as estimates since these were determined using the principle of least squares.

Expanding the expression above for $\hat{\beta}$, we find that it can be written as

$$\hat{\beta} = \frac{n\sum_{i=1}^{n}x_iy_i - \left(\sum_{i=1}^{n}x_i\right)\left(\sum_{i=1}^{n}y_i\right)}{n\sum_{i=1}^{n}x_i^2 - \left(\sum_{i=1}^{n}x_i\right)^2}$$

In our example, we find

$$\hat{\beta} = \frac{5(40155) - (275)(705)}{5(16125) - (275)^2} = 1.38$$

and

$$\hat{\alpha} = 141 - 1.38(55) = 65.1$$

The expressions for $\hat{\alpha}$ and $\hat{\beta}$ are called *least squares estimates* and the straight line they produce is called the *least squares regression line*.

For a given value of x, say x_i, the predicted value for y_i, $\hat{y}_i$ is

$$\hat{y}_i = \hat{\alpha} + \hat{\beta}x_i$$

Now we have equations that can be used with any data set, although a computer is of great value for a large data set. Many statistical computer programs (such as Minitab) produce the least squares line from a data set.

INFLUENTIAL OBSERVATIONS

It turns out that the principle of least squares does not treat all the observations equally. We investigate to see why this is so.

Since $\sum_{i=1}^{n}(x_i - \bar{x})^2$ is constant for a given data set and

$$\hat{\beta} = \frac{\sum_{i=1}^{n}(x_i - \bar{x})(y_i - \bar{y})}{\sum_{i=1}^{n}(x_i - \bar{x})^2}$$

we can write

$$\hat{\beta} = \sum_{i=1}^{n}\frac{(x_i - \bar{x})(y_i - \bar{y})}{\sum_{i=1}^{n}(x_i - \bar{x})^2}$$

So assuming $a_i = (x_i - \bar{x})/\sum_{i=1}^{n}(x_i - \bar{x})^2$, our formula for $\hat{\beta}$ becomes

$$\hat{\beta} = \sum_{i=1}^{n} a_i(y_i - \bar{y}) \quad \text{where} \quad a_i = \frac{(x_i - \bar{x})}{\sum_{i=1}^{n}(x_i - \bar{x})^2}$$

Now the value of a_i depends on the value of x_i, so it appears that $\hat{\beta}$ is a weighted sum of the deviations $y_i - \bar{y}$. But this expression for $\hat{\beta}$ can be simplified even further. Now $\hat{\beta} = \sum_{i=1}^{n} a_i(y_i - \bar{y})$, and this can be written as $\hat{\beta} = \sum_{i=1}^{n} a_i y_i - \bar{y}\sum_{i=1}^{n} a_i$. Now notice that

$$\sum_{i=1}^{n} a_i = \sum_{i=1}^{n}\frac{(x_i - \bar{x})}{\sum_{i=1}^{n}(x_i - \bar{x})^2} = \frac{1}{\sum_{i=1}^{n}(x_i - \bar{x})^2}\sum_{i=1}^{n}(x_i - \bar{x}) = 0$$

since $\sum_{i=1}^{n}(x_i - \bar{x}) = 0$.
So

$$\hat{\beta} = \sum_{i=1}^{n} a_i y_i \quad \text{where} \quad a_i = \frac{(x_i - \bar{x})}{\sum_{i=1}^{n}(x_i - \bar{x})^2}.$$

This shows that the least squares estimate for the slope, $\hat{\beta}$, is a weighted average of the y values. Notice that the values for a_i highly depend upon the value for $x_i - \bar{x}$; the farther x_i is from $\bar{x}$, the larger or smaller is a_i. This is why we pointed out that least squares does not treat all the data points equally. We also note now that if for some point $x_i = \bar{x}$, then that point has absolutely no influence whatsoever on $\hat{\beta}$. This is a fact to which we will return when we consider another straight line to fit a data set, called the *median–median* line.

To continue with our example, we find that $\sum_{i=1}^{n}(x_i - \bar{x})^2 = 1000$, giving the following values for a_i:

$$a_1 = \frac{(35 - 55)}{1000} = -\frac{1}{50}$$

$$a_2 = \frac{(45 - 55)}{1000} = -\frac{1}{100}$$

$$a_3 = \frac{(55 - 55)}{1000} = 0$$

$$a_4 = \frac{(65 - 55)}{1000} = \frac{1}{100}$$

$$a_5 = \frac{(75 - 55)}{1000} = \frac{1}{50}$$

Note that $\sum_{i=1}^{n} a_i = 0$, as we previously noted.
So

$$\hat{\beta} = \sum_{i=1}^{n} a_i y_i = -\frac{1}{50} \cdot 114 - \frac{1}{100} \cdot 124 + 0 \cdot 143 + \frac{1}{100} \cdot 158 + \frac{1}{50} \cdot 166$$

$$= 1.38.$$

Then $\hat{\alpha} = \bar{y} - \hat{\beta}\bar{x} = 141 - 1.38(55) = 65.1$ as before.

We now present a fairly large data set. As we shall see, unlike our little blood pressure example, dealing with it presents some practical as well as some mathematical difficulties.

THE INDY 500

We now turn to a much larger data set, namely, the winning speeds at the Indianapolis 500-mile automobile race conducted every spring. The first race occurred in 1911 and since then it has been held every year, except 1917 and 1918 and 1942–1946 when the race was suspended due to World War I and World War II, respectively. The data are provided in Table 13.3.

A graph of the data, produced by the computer algebra program Mathematica, is shown in Figure 13.3.

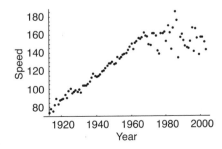

Figure 13.3

Table 13.3

Year	Speed	Year	Speed	Year	Speed
1911	74.602	1944	*	1976	148.725
1912	78.719	1945	*	1977	161.331
1913	75.933	1946	*	1978	161.363
1914	82.474	1947	116.338	1979	158.899
1915	89.840	1948	119.814	1980	142.862
1916	84.001	1949	121.327	1981	139.084
1917	*			1982	162.029
1918	*	1950	124.002	1983	162.117
1919	88.050	1951	126.244	1984	163.612
1920	88.618	1952	128.922	1985	152.982
1921	89.621	1953	128.740	1986	170.722
1922	94.484	1954	130.840	1987	162.175
1923	90.954	1955	128.213	1988	144.809
1924	98.234	1956	128.490	1989	167.581
1925	101.127	1957	135.601	1990	185.981
1926	95.904	1958	133.791	1991	176.457
1927	97.545	1959	135.857	1992	134.477
1928	99.482	1960	138.767	1993	157.207
1929	97.585	1961	139.130	1994	160.872
1930	100.448	1962	140.293	1995	153.616
1931	96.629	1963	143.137	1996	147.956
1932	104.144	1964	147.350	1997	145.827
1933	104.162	1965	150.686	1998	145.155
1934	104.863	1966	144.317	1999	153.176
1935	106.240	1967	151.207	2000	167.607
1936	109.069	1968	152.882	2001	141.574
1937	113.580	1969	156.867	2002	166.499
1938	117.200	1970	155.749	2003	156.291
1939	115.035	1971	157.735	2004	138.518
1940	114.277	1972	162.962	2005	157.603
1941	115.117	1973	159.036	2006	157.085
1942	*	1974	158.589	2007	151.774
1943	*	1975	149.213	2008	143.562

The least squares regression line is Speed $= -1649.13 + 0.908574 \times$ Year. The calculations for determining $\hat{\alpha}$ and $\hat{\beta}$ are quite formidable for this (and almost any other real) data set. The calculations and many of the graphs in this chapter were made by the statistical program Minitab or the computer algebra system Mathematica.

In Figure 13.3, one can notice the years in which the race was not held and the fact that the data appear to be linear until 1970 or so when the winning speeds appear to become quite variable and scattered. We will consider the data in three parts, using the partitions the war years provide. We will also use these partitions when we consider another line to fit to the data, called the *median–median* line. Some of the statistics for the data are shown in Table 13.4.

Table 13.4

Years	Mean	Median	Variance
1911–1916	80.93	80.60	32.22
1919–1941	101.84	100.45	80.78
1947–2008	148.48	149.95	214.63

Clearly, the speeds during 1947–2008 have not only increased the mean winning speed but also had a large influence on the variance of the speeds. We will return to this discussion later. The principle of least squares can be used with any data set, no matter whether it is truly linear or not. So we begin our discussion on whether the line fits the data well or not.

A TEST FOR LINEARITY: THE ANALYSIS OF VARIANCE

It is interesting to fit a straight line to a data set, but the line may or may not fit the data well. One could take points, for example, on the boundary of a semicircle and not get a fit that was at all satisfactory. So we seek a test or procedure that will give us some information on how well the line fits the data.

Is our regression line a good approximation to the data? The answer depends on the accuracy one desires in the line. A good idea, always, is to compare the observed values of y with the values predicted by the line. In the following table, let $\hat{y}$ denote the predicted value for y. If we do this for our blood pressure data set, we find the following values.

x_i	y_i	$\hat{y}_i$
35	114	113.4
45	124	127.2
55	143	141.0
65	158	154.8
75	166	168.6

If the predicted values are sufficient for the experimenter, then the model is a good one and no further investigation is really necessary. If the experimenter wishes to have a measure of the adequacy of the fit of the model, then a test is necessary. We give here what is commonly known as the *analysis of variance.*

First, consider what we called the *total sum of squares*, or SST. This is the sum of squares of the deviations of the y values from their mean, $\bar{y}$:

$$SST = \sum_{i=1}^{n}(y_i - \bar{y})^2$$

For our data set, since $\bar{y} = 141$, this is

$$SST = (114-141)^2 + (124-141)^2 + (143-141)^2 + (158-141)^2 + (166-141)^2$$
$$= 1936.0$$

In using the principle of least squares, we considered the sum of squares of the deviations of the observed y values and their predicted values from the line. We simply called this quantity SS above. We usually call this the *sum of squares due to error*, or SSE, since it measures the discrepancies between the observed and predicted values of y:

$$SSE = \sum_{i=1}^{n}(y_i - \hat{y}_i)^2$$

For our data set, this is

$$SSE = (114-113.4)^2+(124-127.2)^2+(143-141)^2+(158-154.8)^2+(166-168.6)^2$$
$$= 31.6$$

This is of course the quantity we wished to minimize when we used the principle of least squares and this is its minimum value.

Finally, consider SSR or what we call the *sum of squares due to regression*. This is the sum of squares of the deviations between the predicted values, $\hat{y}_i$, and the mean of the y values, $\bar{y}$:

$$SSR = \sum_{i=1}^{n}(\hat{y}_i - \bar{y})^2$$

For our data set, this is

$$SSR = (113.4-141)^2+(127.2-141)^2+(141-141)^2+(154.8-141)^2+(168.6-141)^2$$
$$= 1904.4$$

We notice here that these sums of squares add together as $1904.4 + 31.6 = 1936$ or $SST = SSR + SSE$.

Note that SSR is a large portion of SST. This indicates that the least squares line is a good fit for the data. Now the identity $\text{SST} = \text{SSR} + \text{SSE}$ in this case is not a coincidence; this is always true!

To prove this, consider the identity

$$y_i - \bar{y} = (y_i - \hat{y}_i) + (\hat{y}_i - \bar{y})$$

Now square both sides and sum over all the values giving

$$\sum_{i=1}^{n}(y_i - \bar{y})^2 = \sum_{i=1}^{n}(y_i - \hat{y}_i)^2 + 2\sum_{i-1}^{n}(y_i - \hat{y}_i)(\hat{y}_i - \bar{y}) + \sum_{i=1}^{n}(\hat{y}_i - \bar{y})^2$$

Note for the least squares line,

$$\hat{\beta} = \frac{\sum_{i=1}^{n}(x_i - \bar{x})(y_i - \bar{y})}{\sum_{i=1}^{n}(x_i - \bar{x})^2}$$

and

$$\hat{\alpha} = \bar{y} - \hat{\beta}\bar{x}$$

Now $\hat{y}_i = \hat{\alpha} + \hat{\beta}x_i = \bar{y} - \hat{\beta}\bar{x} + \hat{\beta}x_i = \bar{y} + \hat{\beta}(x_i - \bar{x})$, so $\hat{y}_i - \bar{y} = \hat{\beta}(x_i - \bar{x})$ and the middle term above (ignoring the 2) is

$$\sum_{i-1}^{n}(y_i - \hat{y}_i)(\hat{y}_i - \bar{y}) = \sum_{i-1}^{n}[(y_i - \bar{y}) - \hat{\beta}(x_i - \bar{x})]\hat{\beta}(x_i - \bar{x})$$

$$= \hat{\beta}\sum_{i-1}^{n}(y_i - \bar{y})(x_i - \bar{x}) - \hat{\beta}^2\sum_{i-1}^{n}(x_i - \bar{x})^2$$

$$= 0$$

since

$$\hat{\beta} = \frac{\sum_{i=1}^{n}(x_i - \bar{x})(y_i - \bar{y})}{\sum_{i=1}^{n}(x_i - \bar{x})^2}$$

So,

$$\text{SST} = \sum_{i=1}^{n}(y_i - \bar{y})^2 = \sum_{i=1}^{n}(y_i - \hat{y}_i)^2 + 2\sum_{i-1}^{n}(y_i - \hat{y}_i)(\hat{y}_i - \bar{y}) + \sum_{i=1}^{n}(\hat{y}_i - \bar{y})^2$$

becomes

$$\sum_{i=1}^{n}(y_i - \bar{y})^2 = \sum_{i=1}^{n}(y_i - \hat{y}_i)^2 + \sum_{i=1}^{n}(\hat{y}_i - \bar{y})^2$$

We could say that $\sum_{i=1}^{n}(y_i - \bar{y})^2$ represents the *total sum of squares* of the observations around their mean value. We denote this by SST.

We could also say that $\sum_{i=1}^{n}(y_i - \hat{y}_i)^2$ represents the *sum of squares due to error* or SSE. This is the quantity that the principle of least squares seeks to minimize.

Finally, we could say that $\sum_{i=1}^{n}(\hat{y}_i - \bar{y})^2$ represents the *sum of squares due to regression* or SSR. So the identity above,

$$\sum_{i=1}^{n}(y_i - \bar{y})^2 = \sum_{i=1}^{n}(y_i - \hat{y}_i)^2 + \sum_{i=1}^{n}(\hat{y}_i - \bar{y})^2$$

can be abbreviated as SST = SSE + SSR, as we have seen in our example previously. This partition of the total sum of squares is often called an *analysis of variance* partition, although it has little to do with a variance.

If the data show a strong linear relationship, we expect SSR = $\sum_{i=1}^{n}(\hat{y}_i - \bar{y})^2$ to be large since we expect the deviations $\hat{y}_i - \bar{y}$ to be a large proportion of the total sum of squares $\sum_{i=1}^{n}(y_i - \bar{y})^2$. On the contrary, if the data are actually not linear, we expect the sum of squares due to error, $\sum_{i=1}^{n}(y_i - \hat{y}_i)^2$, to be a large proportion of the total sum of squares $\sum_{i=1}^{n}(y_i - \bar{y})^2$.

This suggests that we look at the ratio SSR/SST, which is called the *coefficient of determination* and is denoted by r^2. Its square root, r, is called the *correlation coefficient*.

For our data, $r^2 = 1904.4/1936 = 0.98368$; so $r = 0.9918$. (The positive square root of r^2 is used if y increases with increasing x while the negative square root of r^2 is used if y decreases with increasing x.)

Since SSR is only a part of SST, it follows that

$$0 \leq r^2 \leq 1$$

so that

$$-1 \leq r \leq 1$$

It would be nice if this number by itself would provide an accurate test for the adequacy of the regression, but unfortunately this is not so. There are data sets for which r is large but the fit is poor and there are data sets for which r is small and the fit is very good.

It is interesting to compare the prewar years (1911–1941) with the postwar years (1947–2008). The results from the analysis of variance are provided in Table 13.5 and the least squares regression lines are provided in Table 13.6.

Table 13.5

Years	SSE	SSR	SST	r^2
1911–1941	8114	4978	13092	0.9396
1947–2008	293	3777	4020	0.3802

Table 13.6

Years	Least Squares Line
1911–1941	Speed $= -2357.99 + 1.27454 \times$ Year
1947–2008	Speed $= -841.651 + 0.500697 \times$ Year

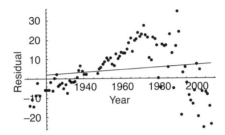

Figure 13.4

From the values for r^2, we conclude that the linear fit is very good for the years 1911–1941 and that the fit is somewhat less good for the years 1947–2008. The graph in Figure 13.4 shows the scatter plot of all the data and the least squares line.

A Caution

Conclusions based on the value of r^2 alone can be very misleading. Consider the following data set.

x	13	4	10	14	9	11	
y	17.372	7.382	9.542	21.2	7.932	11.652	
x	15	7	12	6	12	8	16
y	25.092	6.212	14.262	6.102	0.500	6.822	29.702

The analysis of variance gives $r^2 = 0.59$, which might be regarded as fair but certainly not a large value. The data here have not been presented in increasing values for x, so it is difficult to tell whether the regression is linear or not. The graph in Figure 13.5 persuades us that a linear fit is certainly not appropriate! The lesson here is *always draw a graph*!

NONLINEAR MODELS

The procedures used in simple linear regression can be applied to a variety of nonlinear models by making appropriate transformations of the data. This could be done in the example above, where the relationship is clearly quadratic. For example, if we wish

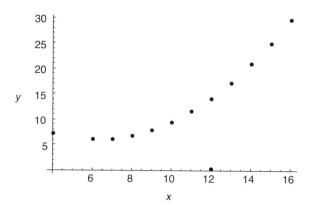

Figure 13.5

to fit a model of the form $y_i = a \cdot 10^{bx_i}$ to a data set, we can take logarithms to find $\log y_i = \log a + bx_i$.

Then our simple linear regression procedure would give us estimates of b and $\log a$. We would then take the antilog to estimate a. Here are some other models, including the quadratic relationship mentioned above, and the appropriate transformations:

1. $y_i = a \cdot 10^{b/x_i}$. Logarithms give $\log y_i = \log a + b/x_i$, so we fit our straight line to the data set $\{1/x_i, \log y_i\}$.

2. $y_i = a \cdot x_i^b$. Taking logarithms gives $\log y_i = \log a + b \log x_i$, so the linear model is fitted to the data set $\{\log x_i, \log y_i\}$.

3. $y_i = a + b \log x_i$ uses the data set $\{\log x_i, y_i\}$.

4. $y_i = 1/(a + b \cdot 10^{-x_i})$ can be transformed into $1/y_i = a + b \cdot 10^{-x_i}$. The linear model is then fitted to the data set $\{10^{-x_i}, 1/y_i\}$.

5. The model $y_i = x_i/(ax_i - b)$ can be transformed into $1/y_i = a - b/x_i$.

THE MEDIAN–MEDIAN LINE

Data sets are often greatly influenced by very large or very small observations; we explored this to some extent in Chapter 12. For example, suppose salaries in a small manufacturing plant are as follows:

$$\$12,500, \ \$13,850, \ \$21,900, \ \$26,200, \ \$65,600$$

The mean of these salaries is $27,410, but four out of five workers receive less than this average salary! The mean is highly influenced by the largest salary. The median salary (the salary in the middle or the mean of the two middlemost salaries when the salaries are arranged in order) is $21,900. We could even replace the two highest salaries with salaries equal to or greater than $21,900, and the median would remain at $21,900 while the mean might be heavily influenced.

For this reason, the median is called a *robust* statistic since it is not influenced by extremely large (or small) values.

The median can also be used in regression. We now describe the *median–median line*. This line is much easier to calculate than the least squares line and enjoys some surprising connections with geometry.

We begin with a general set of data $\{x_i, y_i\}$ for $i = 1, ..., n$. To find the median–median line, first divide the data into three parts (which usually contain roughly the same number of data points). In each part, determine the median of the x-values and the median of the y-values (this will rarely be one of the data points). These points are then plotted as the vertices of a triangle. P_1 contains the smallest x-values, P_2 contains the middle x-values, and P_3 contains the largest x-values.

The median–median line is determined by drawing a line parallel to the baseline (the line joining P_1 and P_3) and at $1/3$ of the distance from the baseline toward P_2. The concept is that since P_1 and P_3 contain $2/3$ of the data and P_2 contains $1/3$ of the data, the line should be moved $1/3$ of the distance from the baseline toward P_2. The general situation is shown in Figure 13.6.

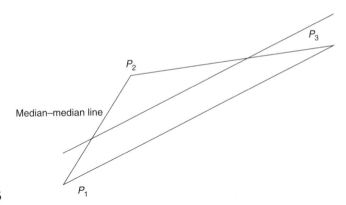

Figure 13.6

Now for some geometric facts.

The medians of a triangle meet at a point that is $1/3$ of the distance from the baseline. Figure 13.7 shows these medians and their meeting point.

If the triangle is determined by the points $P_1(x_1, y_1)$, $P_2(x_2, y_2)$, and $P_3(x_3, y_3)$, then the medians meet at the point $(\bar{x}, \bar{y})$ where $\bar{x} = 1/3(x_1 + x_2 + x_3)$ and $\bar{y} = 1/3(y_1 + y_2 + y_3)$. (The means here refer to the coordinates of the median points, not the data set in general.) It is also true that the meeting point of the medians is $1/3$ of the distance from the baseline. Let us prove these facts before making use of them in determining the median–median line.

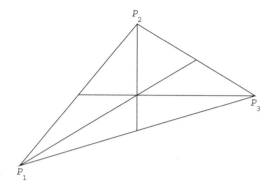

Figure 13.7

Figure 13.8 shows the situation in general. We have taken the baseline along the x-axis, with the leftmost vertex of the triangle at the point (0, 0). This simplifies the calculations greatly and does not limit the generality of our arguments.

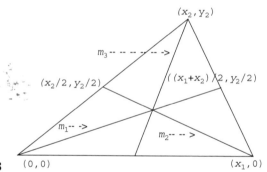

Figure 13.8

First the point $(\bar{x}, \bar{y})$ is the point $((x_1 + x_2)/3, (y_2/3))$, so the point $(\bar{x}, \bar{y})$ is 1/3 of the distance from the baseline toward the vertex, P_2. Now we must show that the medians meet at that point.

The equations of the three medians shown are as follows:

$$m_1: \quad y = \frac{y_2}{x_1 + x_2} . x$$

$$m_2: \quad y = -\frac{x_1 y_2}{x_1 - 2x_2} + \frac{y_2}{x_2 - 2x_1} . x$$

$$m_3: \quad y = -\frac{x_1 y_2}{2x_2 - x_1} + \frac{2y_2}{2x_2 - x_1} . x$$

It is easy to verify that the point $((x_1 + x_2)/3, (y_2/3))$ lies on each of the lines and hence is the point of intersection of these median lines.

These facts suggest two ways to determine the equation of the median–median line. They are as follows:

1. First, determine the slope of the line joining P_1 and P_3. This is the slope of the median–median line. Second, determine the point $(\bar{x}, \bar{y})$. Finally, find the median–median line using the slope and a point on the line.

2. Determine the slope of the line joining P_1 and P_3. Then determine the equations of two of the medians and solve them simultaneously to determine the point of intersection. The median–median line can be found using the slope and a point on the line.

   To these two methods, suggested by the facts above, we add a third method.

3. Determine the *intercept* of the line joining P_1 and P_3 and the intercept of the line through P_2 with the slope of the line through P_1 and P_3. The intercept of the median–median line is the average of twice the first intercept plus the second intercept (and its slope is the slope of the line joining P_1 and P_3).

Method 1 is by far the easiest of the three methods although all are valid. Methods 2 and 3 are probably useful only if one wants to practice finding equations of lines and doing some algebra! Method 2 is simply doing the proof above with actual data. We will not prove that method 3 is valid but the proof is fairly easy.

When Are the Lines Identical?

It turns out that if $x_2 = \bar{x}$, then the least squares line and the median–median line are identical. To show this, consider the diagram in Figure 13.9 where we have taken the base of the triangle along the x-axis and the vertex of the triangle at $(x_1/2, y_2)$ since

$$\bar{x} = \frac{0 + x_1 + \dfrac{x_1}{2}}{3} = \frac{x_1}{2}$$

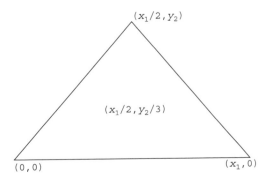

Figure 13.9 $(0,0)$

Now

$$\hat{\beta} = \frac{\sum_{i=1}^{n}(x_i - \bar{x})(y_i - \bar{y})}{\sum_{i=1}^{n}(x_i - \bar{x})^2}$$

Here

$$\sum_{i=1}^{n}(x_i - \bar{x})(y_i - \bar{y}) = \left(0 - \frac{x_1}{2}\right)\left(0 - \frac{y_2}{3}\right) + \left(\frac{x_1}{2} - \frac{x_1}{2}\right)\left(y_2 - \frac{y_2}{3}\right)$$

$$+ \left(x_1 - \frac{x_1}{2}\right)\left(0 - \frac{y_2}{3}\right) = 0$$

so $\hat{\beta} = 0$. Also, $\hat{\alpha} = \bar{y} - \hat{\beta}\bar{x} = \bar{y} = y_2/3$. So the least squares line is $y = y_2/3$. But this is also the median–median line since the median–median line passes through $(\bar{x}, \bar{y})$ and has slope 0.

It is not frequent that $x_2 = \bar{x}$, but if these values are close, we expect the least squares line and the median–median line to also be close.

We now proceed to an example using the Indianapolis 500-mile race winning speeds.

The reason for using these speeds as an example is that the data are divided naturally into three parts due to the fact that the race was not held during World War I (1917 and 1918) and World War II (1942–1946). We admit that the three periods of data are far from equal in size. The data have been given above by year; we now show the median points for each of the three time periods (Table 13.7 and Figure 13.10).

Table 13.7

Period	Median (years)	Median (speed)
1911–1916	1913.5	80.60
1919–1941	1930	100.45
1947–2008	1977.5	149.95

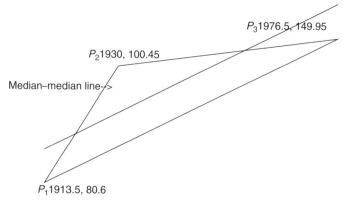

Figure 13.10

Determining the Median–Median Line

The equation of the median–median line will now be determined by each of the three methods described above.

- **Method 1**

    The slope of the median–median line is the slope of the line joining P_1 and P_3.

$$\frac{149.95 - 80.6}{1977.5 - 1913.5} = 1.08359$$

    The point $(\bar{x}, \bar{y})$ is the point

$$\left(\frac{1913.5 + 1930 + 1977.5}{3}, \frac{80.6 + 100.45 + 149.95}{3}\right) = (1940.333, \ 110.333)$$

    So the equation of the median–median line is

$$\frac{y - 110.333}{x - 1940.333} = 1.08359$$

    This can be simplified to $y = -1992.19 + 1.08359\,x$, where y is the speed and x is the year.

- **Method 2**

    We determine the equations of two of the median lines and show that they intersect at the point $(\bar{x}, \bar{y})$. The line from P_1 to the midpoint of the line joining P_2 and P_3 (the point 1973.75, 125.2) is $y = 1.10807x - 2039.69$. The line from P_3 to the midpoint of the line joining P_1 and P_2 (the point 1921.75, 90.525) is $y = 1.06592x - 1957.91$. These lines intersect at (1940.33, 110.33), thus producing the same median–median line as in method 1.

- **Method 3**

    Here we find the intercept of the line joining P_1 and P_3. This is easily found to be -1992.85. Then the intercept of the line through P_2 with slope of the line joining P_1 and P_3(1.08359) is -1990.88. Then weighting the first intercept twice as much as the second intercept, we find the intercept for the median–median line to be

$$\frac{2(-1992.85) + (-1990.88)}{3} = -1992.19$$

    So again we find the same median–median line. The least squares line for the three median points is $y = -2067.73 + 1.12082x$.

These lines appear to be somewhat different, as shown in Figure 13.11. They meet at the point (1958, 129) approximately. It is difficult to compare these lines since the analysis of variance partition

$$\sum_{i=1}^{n}(y_i - \bar{y})^2 = \sum_{i=1}^{n}(y_i - \hat{y}_i)^2 + \sum_{i=1}^{n}(\hat{y}_i - \bar{y})^2$$

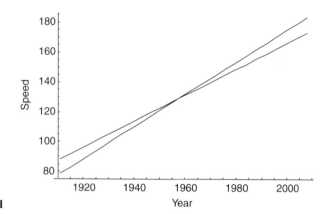

Figure 13.11

or SST $=$ SSE $+$ SSR no longer holds, because the predictions are no longer those given by least squares. For the least squares line, we find

	SST	SSE	SSR	r^2
Least squares line	68,428	12323	56105	0.819918

It is true, however, that

$$\sum_{i=1}^{n}(y_i - \bar{y})^2 = \sum_{i=1}^{n}(y_i - \hat{y}_i)^2 + 2\sum_{i-1}^{n}(y_i - \hat{y}_i)(\hat{y}_i - \bar{y}) + \sum_{i=1}^{n}(\hat{y}_i - \bar{y})^2$$

The total sum of squares remains at $\sum_{i=1}^{n}(y_i - \bar{y})^2 = 68,428$, but the middle term is no longer zero.

We find, in fact, that in this case $\sum_{i=1}^{n}(y_i - \hat{y}_i)^2 = 75933$, $2\sum_{i=1}^{n}(y_i - \hat{y}_i)(\hat{y}_i - \bar{y}) = -25040$, and $\sum_{i=1}^{n}(\hat{y}_i - \bar{y})^2 = 17535$, so the middle term has considerable influence.

There are huge residuals from the predictions using either line, especially in the later years. However, Figure 13.3 shows that speeds become very variable and apparently deviate greatly from a possible straight line relationship during 1911–1970.

We can calculate all the residuals from both the median–median line and the least squares line. Plots of these are difficult to compare. We show these in Figures 13.12 and 13.13.

It is not clear what causes this deviance from a straight line in these years, but in 1972 wings were allowed on the cars, making the aerodynamic design of the car of greater importance than the power of the car. In 1974, the amount of gasoline a car could carry was limited, producing more pit stops; practice time was also reduced.

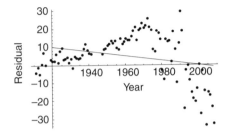

Figure 13.12   Median–median line residuals.

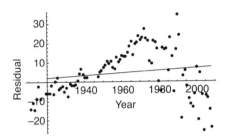

Figure 13.13   Least squares residuals.

Both of these measures were taken to preserve fuel. For almost all the races, some time is spent under the yellow flag. At present, all cars must follow a pace car and are not allowed to increase their speed or pass other cars. Since the time under the yellow flag is variable, this is no doubt a cause of some of the variability of the speeds in the later years. It is possible to account for the time spent on the race under the yellow flag, but that discussion is beyond our scope here. The interested reader should consult a reference on the analysis of covariance.

These considerations prompt an examination of the speeds from the early years only. We have selected the period 1911–1969.

ANALYSIS FOR YEARS 1911–1969

For these years, we find the least squares regression line to be Speed $= -2315.89 + 1.2544\times$ Year and $r^2 = 0.980022$, a remarkable fit.

For the median–median line, we use the points $(1913.5, 80.60)$, $(1930, 100.45)$, and $(1958, 135.601)$, the last point being the median point for the years 1947 through 2008. We find the median–median line to be $y = 1.23598x - 2284.63$. The lines are very closely parallel, but have slightly different intercepts. Predictions based upon them will be very close. These lines are shown in Figure 13.14.

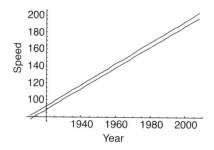

Figure 13.14

CONCLUSIONS

The data from the winning speeds at the Indianapolis 500-mile race provide a fairly realistic exercise when one is confronted with a genuine set of data. Things rarely work out as well as they do with textbook cases of arranged or altered data sets.

We find that in this case, the median–median line is a fine approximation of the data for the early years of the data and that the least squares line is also fine for the early years of the data; neither is acceptable for the later years when we speculate that alterations in the aerodynamics of the cars and time spent under the yellow flag produce speeds that vary considerably from a straight line prediction.

EXPLORATIONS

1. Using the three median points for the Indy 500 data, show that method 3 is a valid procedure for finding the median–median line.

2. Using the three median points for the Indy 500 data, find the least squares line for these points.

3. Find the analysis of variance partition for the least squares line in Exploration 2.

4. Analyze the Indy 500 data for 1998–2008 by finding both the median–median and the least squares lines. Show the partitions of the total sum of squares, SST, in each case.

Chapter 14

Sampling

CHAPTER OBJECTIVES:

- to show some properties of simple random sampling
- to introduce stratified sampling
- to find some properties of stratified sampling
- to see how proportional allocation works
- to discuss optimal allocation
- to find some properties of proportional and optimal allocation.

One of the primary reasons that statistics has become of great importance in science and engineering is the knowledge we now have concerning sampling and the conclusions that can be drawn from samples. It is perhaps a curious and counterintuitive fact that knowledge about a population or group can be found with great accuracy by examining a *sample*—only part of the population or group.

Almost all introductory courses in statistics discuss only simple random sampling. In simple random sampling every item in the population is given an equal chance of occurring in the sample, so every item in the population is treated exactly equally. It may come as a surprise to learn that simple random sampling can often be improved upon; that is, other sampling procedures may well be more efficient in providing information about the population from which the sample is selected. In these procedures, not all the sampled items are treated equally! In addition to simple random sampling, we will discuss *stratified sampling* and both *proportional allocation* and *optimal allocation* within stratified sampling.

We start with a very small example so that ideas become clear.

A Probability and Statistics Companion, John J. Kinney
Copyright © 2009 by John Wiley & Sons, Inc.

EXAMPLE 14.1   *High and Middle Schools*

An urban school district is interested in discovering some characteristics of some of its high and middle schools. We emphasize from the beginning that we assume we know the entire population. In practice, however, we would never know this (or else sampling is at best an idle exercise). The details of the population are given in Table 14.1.

Table 14.1

School enrollment	School type
1667	High
2002	High
1493	High
1802	High
1535	High
731	Middle
834	Middle
699	Middle

The mean enrollment is $\mu_x = 1345.375$ and the standard deviation is $\sigma_x = 481.8737$. We want to show how these statistics can be estimated by taking a sample of the schools. The subscript x is used to distinguish the population from the samples we will select. We first consider simple random sampling.   ■

SIMPLE RANDOM SAMPLING

In simple random sampling, each item in the population, in this case a school, has an equal chance of appearing in the sample as any of the other schools.

We decide to choose five of the eight schools as a random sample. If we give each of the eight schools equal probability of occurring in the sample, then we have a *simple random sample*. There are $\binom{8}{5} = 56$ simple random samples. These are shown in Table 14.2.

We note here that the sample size 5 is relatively very large compared to the size of the population, 8, but this serves our illustrative purposes. In many cases, the sample size can be surprisingly small in relative to the size of the population. We cannot discuss sample size here, but refer the reader to many technical works on the subject.

Note that each of the schools occurs in $\binom{7}{4} = 35$ of the random samples, so each of the schools is treated equally in the sampling process.

Since we are interested in the mean enrollment, we find the mean of each of the samples to find the 56 mean enrollments given in Table 14.3.

Call these values of $\bar{x}$. The mean of these values is $\mu_{\bar{x}} = 1345.375$ with standard deviation $\sigma_{\bar{x}} = 141.078$. We see that the mean of the means is the mean of the population. So the sample mean, $\bar{x}$, is an *unbiased estimator* of the true population mean,

Table 14.2

{1667, 2002, 1493, 1802, 1535}, {1667, 2002, 1493, 1802, 731}, {1667, 2002, 1493, 1802, 834},
{1667, 2002, 1493, 1802, 699}, {1667, 2002, 1493, 1535, 731}, {1667, 2002, 1493, 1535, 834}
{1667, 2002, 1493, 1535, 699}, {1667, 2002, 1493, 731, 834},{1667, 2002, 1493, 731, 699},
{1667, 2002, 1493, 834, 699}, {1667, 2002, 1802, 1535, 731}, {1667, 2002, 1802, 1535, 834},
{1667, 2002, 1802, 1535, 699}, {1667, 2002, 1802, 731, 834}, {1667, 2002, 1802, 731, 699},
{1667, 2002, 1802, 834, 699}, {1667, 2002, 1535, 731, 834}, {1667, 2002, 1535, 731, 699}
{1667, 2002, 1535, 834, 699}, {1667, 2002, 731, 834, 699}, {1667, 1493, 1802, 1535, 731},
{1667, 1493, 1802, 1535, 834}{1667, 1493, 1802, 1535, 699}, {1667, 1493, 1802, 731, 834}
{1667, 1493, 1802, 731, 699}, {1667, 1493, 1802, 834, 699}, {1667, 1493, 1535, 731, 834}
{1667, 1493, 1535, 731, 699}, {1667, 1493, 1535, 834, 699}, {1667, 1493, 731, 834, 699}
{1667, 1802, 1535, 731, 834}, {1667, 1802, 1535, 731, 699}, {1667, 1802, 1535, 834, 699},
{1667, 1802, 731, 834, 699}, {1667, 1535, 731, 834, 699}, {2002, 1493, 1802, 1535, 731},
{2002, 1493, 1802, 1535, 834}, {2002, 1493, 1802, 1535, 699}, {2002, 1493, 1802, 731, 834},
{2002, 1493, 1802, 731, 699}, {2002, 1493, 1802, 834, 699}, {2002, 1493, 1535, 731, 834},
{2002, 1493, 1535, 731, 699}, {2002, 1493, 1535, 834, 699}, {2002, 1493, 731, 834, 699},
{2002, 1802, 1535, 731, 834}, {2002, 1802, 1535, 731, 699}, {2002, 1802, 1535, 834, 699},
{2002, 1802, 731, 834, 699}, {2002, 1535, 731, 834, 699}, {1493, 1802, 1535, 731, 834},
{1493, 1802, 1535, 731, 699}, {1493, 1802, 1535, 834, 699}, {1493, 1802, 731, 834, 699}
{1493, 1535, 731, 834, 699}, {1802, 1535, 731, 834, 699}

Table 14.3

{1699.8, 1539, 1559.6, 1532.6, 1485.6, 1506.2, 1479.2, 1345.4, 1318.4, 1339, 1547.4,
1568, 1541, 1407.2, 1380.2, 1400.8, 1353.8, 1326.8, 1347.4, 1186.6, 1445.6, 1466.2,
1439.2, 1305.4, 1278.4, 1299, 1252, 1225, 1245.6, 1084.8, 1313.8, 1286.8, 1307.4,
1146.6, 1093.2, 1512.6, 1533.2, 1506.2, 1372.4, 1345.4, 1366, 1319, 1292, 1312.6,
1151.8, 1380.8, 1353.8, 1374.4, 1213.6, 1160.2, 1279, 1252, 1272.6, 1111.8,
1058.4, 1120.2}

μ_x. So we write

$$E(\overline{x}) = \mu_x$$

Also, the process of averaging has reduced the standard deviation considerably. These values of $\overline{x}$ are best seen in the graph shown in Figure 14.1. Despite the fact that the population is flat, that is, each enrollment occurs exactly once, the graph of the means begins to resemble the normal curve. This is a consequence of the *central limit theorem*.

We must be cautious, however, in calculating the standard deviation of the means. From the central limit theorem, one might expect this to be σ_x^2/n, where σ_x^2 is the variance of the population. This is not exactly so in this case. The reason for this is that we are sampling without replacement from a finite population. The central limit theorem deals with samples from an infinite population. If we are selecting a sample

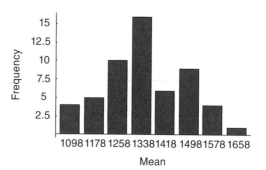

Figure 14.1

of size n from a finite population of size N, then the variance of the sample mean is given by

$$\sigma_{\bar{x}}^2 = \frac{N-n}{N-1} \cdot \frac{\sigma_x^2}{n}$$

where σ_x^2 is the variance of the finite population. Note that in calculating this, we must also find

$$\sigma_x^2 = \sum_{i=1}^{N} \frac{(x_i - \mu)^2}{N}$$

where μ is the mean of the population. Note the divisor of N since we are not dealing with a sample, but with the entire population. Many statistical computer programs assume that the data used are that of a sample, rather than that of a population and use a divisor of $N-1$. So the variance of the population calculated by using a computer program must be multiplied by $(N-1)/N$ to find the correct result.

Here

$$\sigma_{\bar{x}}^2 = \frac{N-n}{N-1} \cdot \frac{\sigma_x^2}{n} = \frac{8-5}{8-1} \cdot \frac{481.87367^2}{5} = 19903.0486$$

and so

$$\sigma_{\bar{x}} = \sqrt{19903.0486} = 141.078$$

exactly the value we calculated using all the sample means.

The factor $(N-n)/(N-1)$ is often called the *finite population correction factor*.

STRATIFICATION

At first glance it might appear that it is impossible to do better than simple random sampling where each of the items in the population is given the same chance as any other item in the population in appearing in the sample. This, however, is not the case! The reason for this is that the population is divided into two recognizable groups, high schools and middle schools. These groups are called *strata*, and in addition to occurring in unequal numbers, they have quite different characteristics. Table 14.4 shows some data from these strata.

Table 14.4

Stratum	Number	Mean	Standard deviation
High Schools	5	1699.800	185.887
Middle Schools	3	754.667	57.5982

The strata then differ markedly in mean values, but, more importantly as we shall see, they vary considerably in variability. We can capitalize on these differences and compose a sampling procedure that is unbiased but has a much smaller standard deviation than that in simple random sampling.

Stratified random sampling takes these different characteristics into account. It turns out that it is most efficient to take simple random samples within each stratum. The question is how to determine the sample size, or allocations, within each stratum. We will show two ways to allocate the sample sizes, called *proportional allocation* and *optimal allocation*, respectively.

Proportional Allocation

In proportional allocation, we choose samples of sizes n_1 and n_2 from strata of sizes N_1 and N_2 so that the proportions in the sample reflect exactly the proportion in the population. That is, we want

$$\frac{n_1}{n_2} = \frac{N_1}{N_2}$$

Since $n = n_1 + n_2$, it follows that $n_1 \cdot N_2 = n_2 \cdot N_1 = (n - n_1) \cdot N_1$ and from this it follows that

$$n_1 = n \cdot \frac{N_1}{N_1 + N_2}$$

and so

$$n_2 = n \cdot \frac{N_2}{N_1 + N_2}$$

In proportional allocation, the sample sizes taken in each stratum are then proportional to the sizes of each stratum. In this case, keeping a total sample of size 5, we take the proportion $5/(5 + 3) = 5/8$ from the high school stratum and $3/(5 + 3) = 3/8$ from the middle school stratum; so we take $5 \cdot 5/8 = 3.125$ observations from the high school stratum and $5 \cdot 3/8 = 1.875$ observations from the middle school stratum. We cannot do this exactly of course, so we choose three items from the high school stratum and two items from the middle school stratum. The sampling within each stratum must be done using simple random sampling. We found $\binom{5}{3} \cdot \binom{3}{2} = 30$ different samples, which are shown in Table 14.5.

Table 14.5

{1667, 2002, 1493, 731, 834}, {1667, 2002, 1493, 731, 699}, {1667, 2002, 1493, 834, 699},
{1667, 2002, 1802, 731, 834}, {1667, 2002, 1802, 731, 699}, {1667, 2002, 1802, 834, 699},
{1667, 2002, 1535, 731, 834}, {1667, 2002, 1535, 731, 699}, {1667, 2002, 1535, 834, 699},
{1667, 1493, 1802, 731, 834}, {1667, 1493, 1802, 731, 699}, {1667, 1493, 1802, 834, 699},
{1667, 1493, 1535, 731, 834}, {1667, 1493, 1535, 731, 699}, {1667, 1493, 1535, 834, 699},
{1667, 1802, 1535, 731, 834}, {1667, 1802, 1535, 731, 699}, {1667, 1802, 1535, 834, 699},
{2002, 1493, 1802, 731, 834}, {2002, 1493, 1802, 731, 699}, {2002, 1493, 1802, 834, 699},
{2002, 1493, 1535, 731, 834}, {2002, 1493, 1535, 731, 699}, {2002, 1493, 1535, 834, 699},
{2002, 1802, 1535, 731, 834}, {2002, 1802, 1535, 731, 699}, {2002, 1802, 1535, 834, 699},
{1493, 1802, 1535, 731, 834}, {1493, 1802, 1535, 731, 699}, {1493, 1802, 1535, 834, 699}

Now we might be tempted to calculate the mean of each of these samples and proceed with that set of 30 numbers. However, this will give a biased estimate of the population mean since the observations in the high school stratum were given 3/2 the probability of appearing in the sample as the observations in the middle school stratum. To fix this, we weight the *mean* of the three observations in the high school stratum with a factor of 5 (the size of the high school stratum) and the *mean* of the two observations from the middle school stratum with a weight of 3 (the size of the high school stratum) and then divide the result by 8 to find the weighted mean of each sample. For example, for the first sample, we find the weighted mean to be

$$\frac{5 \cdot \dfrac{(1667 + 2002 + 1493)}{3} + 3 \cdot \dfrac{(731 + 834)}{2}}{8} = 1368.85$$

which differs somewhat from 1345.40, which is the unweighted mean of the first sample. The set of weighted means is shown in Table 14.6.

A graph of these means is shown in Figure 14.2.

This set of means has mean 1345.375, the true mean of the population; so this estimate for the population mean is also unbiased. But the real gain here is in the standard deviation. The standard deviation of this set of means is 48.6441. This is a large reduction from 141.078, the standard deviation of the set of simple random samples.

Table 14.6

{1368.85, 1343.54, 1362.85, 1433.23, 1407.92, 1427.23, 1377.6, 1352.29, 1371.6,
1327.19, 1301.88, 1321.19, 1271.56, 1246.25, 1265.56, 1335.94, 1310.63, 1329.94,
1396.98, 1371.67, 1390.98, 1341.35, 1316.04, 1335.35, 1405.73, 1380.42, 1399.73,
1299.69, 1274.38, 1293.69}

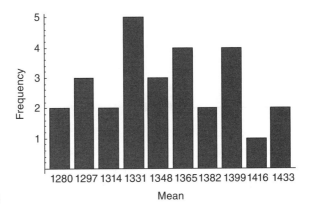

Figure 14.2

This procedure has then cut the standard deviation by about 66% while remaining unbiased. Clearly, stratified sampling has resulted in much greater efficiency in estimating the population mean. One of the reasons for this is the difference in the standard deviations within the strata. The high school stratum has standard deviation 185.887, whereas the middle school stratum has a much smaller standard deviation, 57.5982. This discrepancy can be utilized further in stratification known as *optimal allocation*

Optimal Allocation

We now describe another way of allocating the observations between the strata. The very name, *optimal allocation*, indicates that this is in some sense the best allocation we can devise. We will see that this is usually so and that the standard deviation of the means created this way is even less than that for proportional allocation.

Optimal allocation is derived, in principle, from proportional allocation in which strata with large standard deviations are sampled more frequently than those with smaller standard deviations.

Suppose now that we have two strata: first stratum of size N_1 and standard deviation σ_1 and second stratum of size N_2 and standard deviation σ_2. If the total size of the sample to be selected is n, where n_1 items are selected from the first stratum and n_2 items from the second stratum, then $n = n_1 + n_2$, where these samples sizes are determined so that

$$\frac{n_1}{n_2} = \frac{N_1}{N_2} \cdot \frac{\sigma_1}{\sigma_2}$$

Here the population proportion, N_1/N_2, is weighted by the ratio of the standard deviations, σ_1/σ_2.

Since $n_2 = n - n_1$, we have

$$n_1 N_2 \sigma_2 = n_2 N_1 \sigma_1 = (n - n_1) N_1 \sigma_1$$

and this means that we choose

$$n_1 = n \cdot \frac{N_1 \sigma_1}{N_1 \sigma_1 + N_2 \sigma_2}$$

items from the first stratum and

$$n_2 = n \cdot \frac{N_2 \sigma_2}{N_1 \sigma_1 + N_2 \sigma_2}$$

items from the second stratum.

In this case, then we would select

$$5 \cdot \frac{5(185.887)}{5(185.887) + 3(57.5982)} = 4.216$$

items from the first (high school) stratum and

$$5 \cdot \frac{3(57.5982)}{5(185.887) + 3(57.5982)} = 0.784$$

items from the second (middle school) stratum.

The best we can then do is to select four items from the high school stratum and one item from the middle school stratum. This gives $\binom{5}{4} \cdot \binom{3}{1} = 15$ possible samples, which are shown in Table 14.7.

Table 14.7

{1667, 2002, 1493, 1802, 731}, {1667, 2002, 1493, 1802, 834}, {1667, 2002, 1493, 1802, 699}, {1667, 2002, 1493, 1535, 731}, {1667, 2002, 1493, 1535, 834}, {1667, 2002, 1493, 1535, 699}, {1667, 2002, 1802, 1535, 731}, {1667, 2002, 1802, 1535, 834}, {1667, 2002, 1802, 1535, 699} {1667, 1493, 1802, 1535, 731}, {1667, 1493, 1802, 1535, 834}, {1667, 1493, 1802, 1535, 699}, {2002, 1493, 1802, 1535, 731}, {2002, 1493, 1802, 1535, 834}, {2002, 1493, 1802, 1535, 699}

Now again, as we did with proportional allocation, we do not simply calculate the mean of each of these samples but instead calculate a weighted mean that reflects the differing probabilities with which the observations have been collected. We weight the mean of the high school observations with a factor of 5 (the size of the stratum) while we weight the observation from the middle school stratum with its size, 3, before dividing by 8. For example, for the first sample, the weighted mean becomes

$$\frac{5 \cdot \dfrac{(1667 + 2002 + 1493 + 1802)}{4} + 3 \cdot 731}{8} = 1362.25$$

The complete set of weighted means is

{1362.25, 1400.88, 1350.25, 1320.53, 1359.16, 1308.53, 1368.81, 1407.44,

1356.81, 1289.28, 1327.91, 1277.28, 1341.63, 1380.25, 1329.63}

A graph of these means is shown in Figure 14.3.

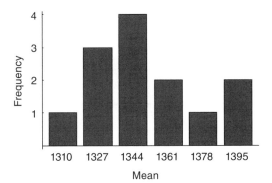

Figure 14.3

The mean of these weighted means is 1345.375, the mean of the population, so once more our estimate is unbiased, but the standard deviation is 36.1958, a reduction of about 26% from that for proportional allocation.

We summarize the results of the three sampling plans discussed here in Table 14.8.

Table 14.8

Sampling	Number	Mean	Standard deviation
Population	8	1345.375	481.874
Simple random	56	1345.375	141.078
Proportional allocation	30	1345.375	48.6441
Optimal allocation	15	1345.375	36.1958

SOME PRACTICAL CONSIDERATIONS

Our example here contains two strata, but in practice one could have many more. Suppose the strata are as follows:

Stratum	Number	Standard deviation
1	N_1	σ_1
2	N_2	σ_2
$\vdots$	$\vdots$	
k	N_k	σ_k

Suppose $N_1 + N_2 + \cdots + N_k = N$ and that we wish to select a sample of size n. In proportional allocation, we want the sample sizes in each stratum to reflect the sizes of the stratum, so we want

$$\frac{n_1}{N_1} = \frac{n_2}{N_2} = \cdots = \frac{n_k}{N_k}$$

The solution to this set of equations is to choose $n_i = n \cdot N_i/N$ observations from the ith stratum. Then the ratio of the number of observations in the ith stratum to the number of observations in the jth stratum is

$$\frac{n_i}{n_j} = \frac{n \cdot \dfrac{N_i}{N}}{n \cdot \dfrac{N_j}{N}} = \frac{N_i}{N_j}$$

the ratio of the number of items in stratum i to the number of items in stratum j.

The numbers N_i are usually known, at least approximately, so one can come close to proportional allocation in most cases.

Optimal allocation, however, poses a different problem, because the number of observations per stratum depends on the standard deviations. In the general case, optimal allocation selects

$$n \cdot \frac{N_i \sigma_i}{N_1 \sigma_1 + N_2 \sigma_2 + \cdots + N_k \sigma_k}$$

from the ith stratum. This requires knowledge, or close approximation, to the standard deviations. In the case of two strata, however, we need only to know the ratio of the standard deviations. The number of items to be selected from the first stratum is

$$n \cdot \frac{N_1 \sigma_1}{N_1 \sigma_1 + N_2 \sigma_2} = n \cdot \frac{N_1 \sigma_1 / \sigma_2}{N_1 \sigma_1 / \sigma_2 + N_2}$$

So the ratio provides all the information needed. For example, if we know that $\sigma_1/\sigma_2 = 2$, $N_1 = 10$, and $N_2 = 15$ and if we wish to select a sample of 7, then we select

$$7 \cdot \frac{10 \cdot 2}{10 \cdot 2 + 15} = 4$$

from the first stratum and

$$n \cdot \frac{N_2 \sigma_2}{N_1 \sigma_1 + N_2 \sigma_2} = n \cdot \frac{N_2}{N_1 \sigma_1 / \sigma_2 + N_2} = 7 \cdot \frac{15}{10 \cdot 2 + 15} = 3$$

items from the second stratum. It is of course unusual for these sample sizes to be integers, so we do the best we can. Usually the total sample size is determined by the cost of selecting each item. In the general case, if the ratio of the standard deviations to each other is known, or can be approximated, then an allocation equivalent to, or approximately equal to, optimal allocation can be achieved.

It is clear that stratification, of either variety, reduces the standard deviation and so increases greatly the accuracy with which predictions can be made. It is often the case that proportional and optimal allocation do not differ very much with respect to the reduction in the standard deviation although it can be shown in general that

$$\sigma^2_{\text{Smple random sampling}} \geq \sigma^2_{\text{Proportional allocation}} \geq \sigma^2_{\text{Optimal allocation}}$$

is usually the case. Instances where this inequality does not hold are very unlikely to be encountered in practice.

STRATA

Stratification is usually a very efficient technique in sampling. It is important to realize that the strata must exist in a recognizable form before the sampling is done. The strata are then groups with recognizable features. In political sampling, for example, strata might consist of urban and rural residents, but within these strata we might sample home owners, apartment dwellers, condominium owners, and so on as substrata. In addition, in national political polling, the strata might differ from state to state. In any event, the strata cannot be made up without the group having some known characteristics. Stratified sampling has been known to provide very accurate estimates in elections; generally the outcome is known, except in extremely tight races, well before the polls close and all the votes have been cast!

CONCLUSIONS

This has been a very brief introduction to varieties of samples that can be chosen from a population. Much more is known about sampling and the interested reader is encouraged to sample some of the many specialized texts on sampling techniques.

EXPLORATIONS

1. The data in the following table show the populations of several counties in Colorado, some of them urban and some rural.

Population	County type
14,046	Rural
5,881	Rural
4,511	Rural
9,538	Rural
550,478	Urban
278,231	Urban
380,273	Urban
148,751	Urban
211,272	Urban

(**a**) Show all the simple random samples of size 4 and draw graphs of the sample means and sample standard deviations.

(**b**) Draw stratified samples of size 4 by
(i) proportional allocation;
(ii) optimal allocation.

(**c**) Calculate weighted means for each of the samples in part (b).

(**d**) Discuss the differences in the above sampling plans and make inferences.

Chapter 15

Design of Experiments

CHAPTER OBJECTIVES:

- to learn how planning "what observations have to be taken in an experiment" can greatly improve the efficiency of the experiment
- to consider *interactions* between *factors* studied in an experiment
- to study factorial experiments
- to consider what to do when the effects in an experiment are confounded
- to look at experimental data geometrically
- to encounter some interesting three-dimensional geometry.

A recent book by David Salsburg is titled *The Lady Tasting Tea* and is subtitled *How Statistics Revolutionized Science in the Twentieth Century*. The title refers to a famous experiment conducted by R. A. Fisher. The subtitle makes quite a claim, but it is largely true. Did statistics revolutionize science and if so, how? The answer lies in our discovery of how to decide what observations to make in a scientific experiment. If observations are taken correctly we now know that conclusions can be drawn from them that are not possible with only random observations. It is our knowledge of the planning (or the *design*) of experiments that allows experimenters to carry out *efficient* experiments in the sense that valid conclusions may then be drawn. It is our object here to explore certain designed experiments and to provide an introduction to this subject. Our knowledge of the design of experiments and the design of sample surveys are the two primary reasons for studying statistics; yet, we can only give a limited introduction to either of these topics in our introductory course. We begin with an example.

EXAMPLE 15.1   *Computer Performance*

A study of the performance of a computer is being made. Only two variables (called *factors*) are being considered in the study: Speed (S) and RAM (R). Each of these variables is being studied at two values (called *levels*). The levels of Speed are 133 MHz and 400 MHz and the levels of RAM are 128 MB and 256 MB.

If we study each level of Speed with each level of RAM and if we make one observation for each factor combination, we need to make four observations. The observation or response here is the time the computer takes, in microseconds, to perform a complex calculation. These times are shown in Table 15.1.

Table 15.1

	Speed (MHz)	
RAM (MB)	133	400
128	27	10
256	18	9

What are we to make of the data? By examining the columns, it appears that the time to perform the calculation is decreased by increasing speed and, by examining the rows, it appears that the time to perform the calculation is decreased by increasing RAM. Since the factors are studied together, it may be puzzling to decide exactly what influence each of these factors has alone. It might appear that the factors should be studied separately, but, as we will see later, it is particularly important that the factors be studied together. Studies involving only one factor are called *one-factor-at-a-time* experiments and are rarely performed. One reason for the lack of interest in such experiments is the fact that factors often behave differently when other factors are introduced into the experiment. When this occurs, we say the factors *interact* with each other. It may be very important to detect such interactions, but obviously, we cannot detect such interactions unless the factors are studied together. We will address this subsequently.

Now we proceed to assess the influence of each of the factors. It is customary, since each of these factors is at two levels or values, to code these as -1 and $+1$. The choice of coding will make absolutely no difference whatever in the end. The data are then shown in Table 15.2 with the chosen codings. The overall mean of the data is 16, shown in parentheses.

One way to assess the influence of the factor speed is to compare the mean of the computation times at the $+1$ level with the mean of the computation times at the -1 level. We divide this difference by 2, the distance between the codings -1 and $+1$, to find

$$\text{Effect of Speed} = \frac{1}{2}\left(\frac{10+9}{2} - \frac{27+18}{2}\right) = -6.5$$

Table 15.2

	Speed	
RAM	-1	$+1$
-1	27	10
$+1$	18	9
		(16)

This means that we decrease the computation time on average by 6.5 units as we move from the -1 level to the $+1$ level.

Now, what is the effect of RAM? It would appear that we should compare the mean computation times at the $+1$ level with those at the -1 level, and again divide by 2 to find

$$\text{Effect of RAM} = \frac{1}{2}\left(\frac{18+9}{2} - \frac{27+10}{2}\right) = -2.5$$

This means that we decrease computation times on average by 2.5 units as we move from the -1 level to the $+1$ level.

One more effect can be studied, namely, the *interaction* between the factors Speed and RAM.

In Figure 15.1a, we show the computation times at the two levels of RAM for speed at its two levels. As we go from the -1 level of RAM to the $+1$ level, the computation times change differently for the different levels of speed, producing lines that are not parallel. Since the lines are not quite parallel, this is a sign of a mild interaction between the factors speed and RAM. In Figure 15.1b, we show two factors that have a very high interaction, that is, the performance of one factor heavily depends upon the level of the other factor.

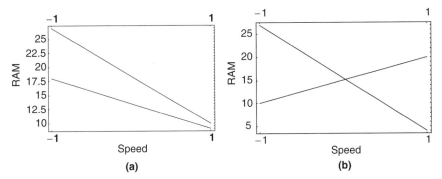

Figure 15.1

The size of the interaction is also a measurement of the effect of the *combination* of levels of the main factors (Speed and RAM). We denote this effect by Speed · RAM. How are we to compute this effect? We do this by comparing the mean where the *product* of the coded signs is $+1$ with the mean where the *product* of the coded signs is -1 to find

$$\text{Effect of speed} \cdot \text{RAM} = \frac{1}{2}\left(\frac{27+9}{2} - \frac{10+18}{2}\right) = 2$$

So we conclude that the computation Speed where the combination of levels of the factors, where the product is $+1$, tends to be two units more in average than the computation Speed where the combination of levels of the factors is -1.

How can we put all this information together? We can use these computed effects in the following model:

$$\text{Observation} = 16 - 6.5 \cdot \text{Speed} - 2.5 \cdot \text{RAM} + 2 \cdot \text{Speed} \cdot \text{RAM}$$

This is called a *linear model*. Linear models are of great importance in statistics, especially in the areas of regression and design of experiments. In this case, we can use the signs of the factors (either $+1$ or -1) in the linear model. For example, if we use speed $= -1$ and RAM $= -1$, we find

$$16 - 6.5(-1) - 2.5(-1) + 2(-1)(-1) = 27$$

exactly the observation at the corner! This also works for the remaining corners in Table 15.2:

$$16 - 6.5(-1) - 2.5(+1) + 2(-1)(+1) = 18$$

$$16 - 6.5(+1) - 2.5(-1) + 2(+1)(-1) = 10$$

and

$$16 - 6.5(+1) - 2.5(+1) + 2(+1)(+1) = 9$$

The linear model then explains exactly each of the observations!   ■

We now explore what happens when another factor is added to the experiment.

EXAMPLE 15.2   *Adding a Factor to the Computer Experiment*

Suppose that we now wish to study two different brands of computers so we add the factor Brand to the experiment. Again, we code the two brands as -1 and $+1$. Now we need a three-dimensional cube to see the computation times resulting from all the combinations of the factors. We show these in Figure 15.2.

Now we have three main effects S, R, and B, in addition to three two-factor interactions, which we abbreviate as SR, SB, and RB, and one three-factor interaction SBR. To calculate

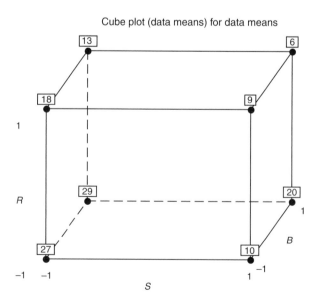

Cube plot (data means) for data means

Figure 15.2

the main effects, we use the planes that form the boundaries of the cube. To find the main S effect, for example, we compare the mean of the plane where S is positive with the mean of the plane where S is negative and take $1/2$ of this difference as usual. This gives

$$S = \frac{1}{2} \left(\frac{9 + 10 + 6 + 20}{4} - \frac{18 + 27 + 13 + 29}{4} \right) = -5.25$$

Similarly, we find

$$R = \frac{1}{2} \left(\frac{6 + 9 + 13 + 18}{4} - \frac{10 + 27 + 29 + 20}{4} \right) = -5$$

and

$$B = \frac{1}{2} \left(\frac{29 + 13 + 6 + 20}{4} - \frac{9 + 10 + 18 + 27}{4} \right) = 0.5$$

The calculation of the interactions now remains. To calculate the SR interaction, it would appear consistent with our previous calculations if we were to compare the plane where SR is positive to the plane where SR is negative. This gives

$$SR = \frac{1}{2} \left(\frac{27 + 9 + 6 + 29}{4} - \frac{18 + 10 + 13 + 20}{4} \right) = 1.25$$

The planes shown in Figure 15.3 may be helpful in visualizing the calculations of the remaining two-factor interactions.
    We find

$$SB = \frac{1}{2} \left(\frac{27 + 18 + 6 + 20}{4} - \frac{10 + 9 + 13 + 29}{4} \right) = 1.25$$

and

$$BR = \frac{1}{2} \left(\frac{10 + 13 + 6 + 27}{4} - \frac{9 + 18 + 20 + 29}{4} \right) = -2.50$$

These two-factor interactions geometrically are equivalent to collapsing the cube along each of its major axes and analyzing the data from the squares that result.

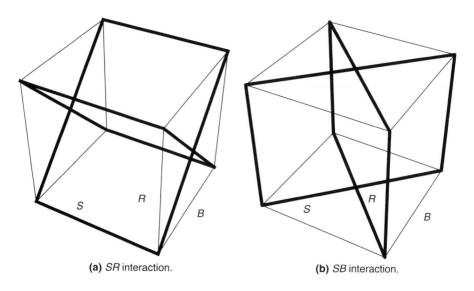

(a) *SR* interaction.   (b) *SB* interaction.

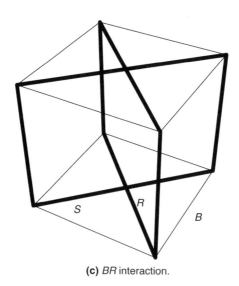

(c) *BR* interaction.

Figure 15.3

This leaves the three-factor interaction *SBR* to be calculated. But we have used up every one of the 12 planes that pass through the cube! Consistent with our previous calculations, if we look at the points where *SBR* is positive, we find a tetrahedron within the cube as shown in Figure 15.4.

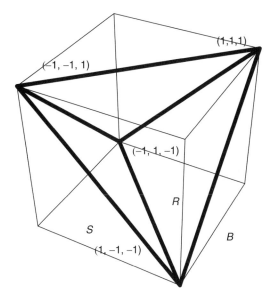

Figure 15.4   Positive tetrahedron.

We also find a negative tetrahedron as shown in Figure 15.5.

Now, we compare the means of the computation times in the positive tetrahedron with that of the negative tetrahedron:

$$SBR = \frac{1}{2}\left(\frac{18 + 10 + 6 + 29}{4} - \frac{27 + 9 + 13 + 20}{4}\right) = -0.75$$

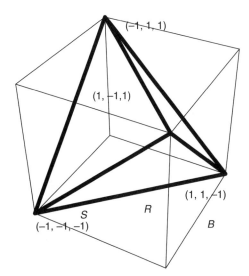

Figure 15.5   Negative tetrahedron.

Now, as in the case of two factors, we form the linear model

$$\text{Observation} = 16.5 - 5.25S - 5R + 0.5B + 1.25SR + 1.25SB - 2.5RB - 0.75SBR$$

where the mean of the eight observations is 16.5.

Again, we find that the model predicts each of the corner observations exactly. We show the signs of the factors (in the order S, R, B) to the left of each calculation.

$$(+, +, +)\quad 16.5 - 5.25 - 5 + 0.5 + 1.25 + 1.25 - 2.5 - 0.75 = 6$$

$$(-, +, +)\quad 16.5 + 5.25 - 5 + 0.5 - 1.25 - 1.25 - 2.5 + 0.75 = 13$$

$$(+, -, +)\quad 16.5 - 5.25 + 5 + 0.5 - 1.25 + 1.25 + 2.5 + 0.75 = 20$$

$$(-, -, +)\quad 16.5 + 5.25 + 5 + 0.5 + 1.25 - 1.25 + 2.5 - 0.75 = 29$$

$$(+, +, -)\quad 16.5 - 5.25 - 5 - 0.5 + 1.25 - 1.25 + 2.5 + 0.75 = 9$$

$$(-, +, -)\quad 16.5 + 5.25 - 5 - 0.5 - 1.25 + 1.25 + 2.5 - 0.75 = 18$$

$$(+, -, -)\quad 16.5 - 5.25 + 5 - 0.5 - 1.25 - 1.25 - 2.5 - 0.75 = 10$$

$$(-, -, -)\quad 16.5 + 5.25 + 5 - 0.5 + 1.25 + 1.25 - 2.5 + 0.75 = 27$$

■

We would like to extend our discussion to four or more factors, but then the nice geometric interpretation we have given in the previous examples becomes impossible. Fortunately, there is another way to calculate the effects that applies to any number of factors. The process is called the *Yates algorithm* after Frank Yates, a famous statistician who discovered it.

YATES ALGORITHM

The calculation of the effects in our examples depends entirely upon the signs of the factors. To calculate the R effect, for example, one would only need to know the sign of R for each of the observations and use the mean of those observations for which R is positive and also the mean of the observations for which the sign of R is negative. So all the effects could be calculated from a table showing the signs of the factors.

It is also possible to calculate all the effects from a table using the Yates algorithm. To do this, the observations must be put in what we will call a natural order. Although the order of the factors is of no importance, we will follow the order S, R, B as we have done previously. We make a column for each of the main factors and make columns of the signs as follows. Under S, make a column starting with the signs $-, +, -, +, \cdots$; under R, make a column with the pattern $-, -, +, +, -, -, +, +, \cdots$; finally under B, make a column $-, -, -, -, +, +, +, +, -, \cdots$. If there are more factors, the next column starts with eight minus signs followed by eight plus signs and this pattern is repeated; the next factor would begin with 16 minus signs followed by 16 plus signs and this pattern continues. The columns are of course long enough to accommodate each of the observations. The result in our example is shown in Table 15.3.

Table 15.3

S	R	B	Observations	1	2	3	Effect ($\div 8$)	
$-$	$-$	$-$	27	37	64	132	16.5	μ
$+$	$-$	$-$	10	27	68	-42	-5.25	S
$-$	$+$	$-$	18	49	-26	-40	-5	R
$+$	$+$	$-$	9	19	-16	10	1.25	SR
$-$	$-$	$+$	29	-17	-10	4	0.50	B
$+$	$-$	$+$	20	-9	-30	10	1.25	SB
$-$	$+$	$+$	13	-9	8	-20	-2.50	RB
$+$	$+$	$+$	6	-7	2	-6	-0.75	SBR

To calculate the effects, proceed as follows. Going down the column of observations, consider the observations in pairs. Add these to find the first entries in the column labeled 1. Here we find $27 + 10 = 37$, $18 + 9 = 27$, $29 + 20 = 49$, and $13 + 6 = 19$. Now consider the same pairs but subtract *the top entry from the bottom entry* to find $10 - 27 = -17$, $9 - 18 = -9$, $20 - 29 = -9$, and $6 - 13 = -7$. This completes column 1.

Now perform exactly the same calculations on the entries in column 1 to find column 2. Finally, follow the same pattern on the entries of column 2 to find the entries in column 3. The effects are found by dividing the entries in column 3 by 8. We find the same model here that we found above using geometry.

Note that the factors described in the rightmost column can be determined from the $+$ signs given in the columns beneath the main factors.

We want to show an example with four factors using the Yates algorithm since the geometry is impossible, but first we make some comments on the experimental design and introduce some notation.

RANDOMIZATION AND SOME NOTATION

Each of our examples are instances of what are called *full factorial experiments*. These experiments make observations for each combination of each level of each factor; no combinations are omitted. In our examples, these are usually denoted by the symbol 2^k, the 2 indicating that each factor is at 2 levels and k indicating the number of factors. Example 15.1 is then a 2^2 experiment while example 15.2 is a 2^3 experiment.

Factorial experiments are very efficient in the sense that although the factors are observed together and would appear to be hopelessly mixed up, they can be shown to be equivalent of one-factor-at-a-time experiments where each factor and their interactions are observed separately 2^k times! Of course, one-factor-at-a-time experiments using only the main effects cannot draw conclusions about interactions since they are never observed.

Both the design and the occurrence of *randomization* are keys to the statistical analysis of an experiment. In factorial experiments, the order of the observations should be determined by some random scheme including repetitive observations for the same factor combinations should these occur.

Now we show an example with four factors.

EXAMPLE 15.3 A 2^4 Factorial Experiment

In *Statistics for Experimenters* by George Box, William Hunter, and J. Stuart Hunter, a chemical process development study is described using four main factors. These are A: catalyst charge (in pounds), B: temperature (in degrees centigrade), C: pressure (in pounds per square inch), and D: concentration (in percentage).

The chemical process is being created and the experimenters want to know what factors and their interactions should be considered when the actual process is implemented. The results are shown in Table 15.4.

Table 15.4

A	B	C	D	Observations	1	2	3	4	Effect ($\div 16$)	
−	−	−	−	71	132	304	600	1156	72.25	μ
+	−	−	−	61	172	296	556	−64	−4	A
−	+	−	−	90	129	283	−32	192	12	B
+	+	−	−	82	167	273	−32	8	0.50	AB
−	−	+	−	68	111	−18	78	−18	−1.125	C
+	−	+	−	61	172	−14	114	6	0.375	AC
−	+	+	−	87	110	−17	2	−10	−0.625	BC
+	+	+	−	80	163	−15	6	−6	−0.375	ABC
−	−	−	+	61	−10	40	−8	−44	−2.75	D
+	−	−	+	50	−8	38	−10	0	0	AD
−	+	−	+	89	−7	61	4	36	2.25	BD
+	+	−	+	83	−7	53	2	4	0.25	ABD
−	−	+	+	59	−11	2	−2	−2	−0.125	CD
+	−	+	+	51	−6	0	−8	−2	−0.125	ACD
−	+	+	+	85	−8	5	−2	−6	−0.375	BCD
+	+	+	+	78	−7	1	−4	−2	−0.125	ABCD

This gives the linear model

$$\text{Observations} = 72.25 - 4A + 12B + 0.50AB - 1.125C + 0.375AC - 0.625BC$$

$$-0.375ABC - 2.75D + 0AD + 2.25BD + 0.25ABD - 0.125CD$$

$$-0.125ACD - 0.375BCD - 0.125ABCD$$

As in the previous examples, the model predicts each of the observations exactly. ∎

CONFOUNDING

It is often true that the effects for the higher order interactions become smaller in absolute value as the number of factors in the interaction increases. So these interactions have something, but often very little, to do with the prediction of the observed values. If it is not necessary to estimate some of these higher order interactions, some very substantial gains can be made in the experiment, for then we do not have to observe all the combinations of factor levels! This then decreases the size, and hence the cost, of the experiment without substantially decreasing the information to be gained from the experiment.

We show this through an example.

EXAMPLE 15.4   *Confounding*

Consider the data in Example 15.3 again, as given in Table 15.4, but suppose that we have only the observations for which $ABCD = +1$. This means that we have only half the data given in Table 15.4. We show these data in Table 15.5, where we have arranged the data in standard order for the three effects A, B, and C (since if the signs for these factors are known, then the sign of D is determined).

Table 15.5

A	B	C	D	Observations	1	2	3	Effect ($\div 8$)	
−	−	−	−	71	121	292	577	72.125	$\mu + ABCD$
+	−	−	+	50	171	285	−35	−4.375	$A + BCD$
−	+	−	+	89	120	−28	95	11.875	$B + ACD$
+	+	−	−	82	165	−7	3	0.375	$AB + CD$
−	−	+	+	59	−21	50	7	0.875	$C + ABD$
+	−	+	−	61	−7	45	21	2.625	$AC + BD$
−	+	+	−	87	2	14	−5	−0.625	$BC + AD$
+	+	+	+	78	−9	−11	−25	−3.125	$ABC + D$

Note that the effects are not the same as those found in Table 15.4. To the mean μ the effect $ABCD$ has been added to find $\mu + ABCD = 72.25 - 0.125 = 72.125$. Similarly, we find $A + BCD = -4 - 0.375 = -4.375$. The other results in Table 15.5 may be checked in a similar way. Since we do not have all the factor level combinations, we would not expect to find the results we found in Table 15.4. The effects have all been somewhat interfered with or *confounded*. However, there is a pattern. To each effect has been added the effect or interaction that is missing from $ABCD$. For example, if we consider AB, then CD is missing and is added to AB. The formula $ABCD = +1$ is called a *generator*.

One could also confound by using the generator $ABCD = -1$. Then one would find $\mu - ABCD$, $A - BCD$, and so on. We will not show the details here.

Note also that if the generator $ABCD = +1$ is used, the experiment is a full factorial experiment in factors A, B, and C.

Each time a factor is added to a full factorial experiment where each factor is at two levels, the number of factor level combinations is doubled, greatly increasing the cost and the time to carry out the experiment. If we can allow the effects to be confounded, then we can decrease the size and the cost of the experiment.

In this example, the confounding divides the size of the experiment in half and so is called a *half fraction* of a 2^4 factorial experiment and is denoted as a 2^{4-1} factorial experiment to distinguish this from a full 2^3 experiment. These experiments are known as *fractional factorial experiments*. These experiments can give experimenters much information with great efficiency provided, of course, that the confounded effects are close to the true effects. This is generally true for confounding using higher order interactions.

Part of the price to be paid here is that all the effects, including the main effects, are confounded. It is generally poor practice to confound main effects with lower order interactions.

If we attempt to confound the experiment described in Example 15.2 by using the generator $SBR = +1$, we find the results given in Table 15.6.

Table 15.6

S	B	R	Observations	(1)	(2)	Effect ($\div 4$)	
−	+	−	29	39	63	15.75	$\mu + SBR$
−	−	+	10	24	−31	−7.75	$S + BR$
+	−	−	18	−19	−15	−3.75	$R + SB$
+	+	+	6	−12	7	1.75	$B + SR$

The problem here is that the main effects are confounded with second-order interactions. This is generally a very poor procedure to follow. Fractional factorial experiments are only useful when the number of factors is fairly large resulting in the main effects being confounded with high-order interactions. These high-order interactions are normally small in absolute value and are very difficult to interpret in any event. When we do have a large number of factors, however, then fractional factorial experiments become very useful and can reduce the size of an experiment in a very dramatic fashion. In that case, multiple generators may define the experimental procedure leading to a variety of confounding patterns. ■

MULTIPLE OBSERVATIONS

We have made only a single observation for each combination of factor levels in each of our examples. In reality, one would make multiple observations whenever possible. This has the effect of increasing the accuracy of the estimation of the effects, but we will not explore that in detail here. We will show an example where we have multiple observations; to do this, we return to Example 15.1, where we studied the effects of the factors Speed and RAM on computer performance. In Table 15.7, we have used three observations for each factor level combination.

Our additional observations happen to leave the means of each cell, as well as the overall mean, unchanged. Now what use is our linear model that was

$$\text{Observation} = 16 - 6.5 \cdot \text{Speed} - 2.5 \cdot \text{RAM} + 2 \cdot \text{Speed} \cdot \text{RAM}?$$

Table 15.7

	Speed	
RAM	−1	+1
−1	27,23,31	10,8,12
+1	18,22,14	9,11,7
		(16)

This linear model predicts the mean for each factor combination but not the individual observations. To predict the individual observations, we add a random component ε to the model to find

$$\text{Observation} = 16 - 6.5 \cdot \text{Speed} - 2.5 \cdot \text{RAM} + 2 \cdot \text{Speed} \cdot \text{RAM} + \varepsilon$$

The values of ε will then vary with the individual observations. For the purpose of statistical inference, which we cannot consider here, it is customary to assume that the random variable ε is normally distributed with mean 0, but a consideration of that belongs in a more advanced course.

DESIGN MODELS AND MULTIPLE REGRESSION MODELS

The linear models developed here are also known as *multiple regression* models. If a multiple regression computer program is used with the data given in any of our examples and the main effects and interactions are used as the independent variables, then the coefficients found here geometrically are exactly the coefficients found using the multiple regression program. The effects found here then are exactly the same as those found using the principle of least squares.

TESTING THE EFFECTS FOR SIGNIFICANCE

We have calculated the effects in factorial designs, and we have examined their size, but we have not determined whether these effects have statistical significance. We show how this is done for the results in Example 15.3, the 2^4 factorial experiment. The size of each effect is shown in Table 15.8.

The effects can be regarded as Student t random variables. To test any of the effects for statistical significance, we must determine the standard error for each effect. To do this, we first determine a set of effects that will be used to calculate this standard error. Here, let us use the third- and fourth-order interactions; we will then test each of the main effects and second-order interactions for statistical significance. We proceed as follows.

Table 15.8

Size	Effect
72.25	μ
−4	A
12	B
0.50	AB
−1.125	C
0.375	AC
−0.625	BC
−0.375	ABC
−2.75	D
0	AD
2.25	BD
0.25	ABD
−0.125	CD
−0.125	ACD
−0.375	BCD
−0.125	$ABCD$

First, find a quantity called a *sum of squares* that is somewhat similar to the sums of squares we used in Chapter 13. This is the sum of the squares of the effects that will be used to determine the standard error. Here this is

$$SS = (ABC)^2 + (ABD)^2 + (ACD)^2 + (BCD)^2 + (ABCD)^2$$
$$= (-0.375)^2 + (0.250)^2 + (-0.125)^2 + (-0.375)^2 + (-0.125)^2$$
$$= 0.375$$

We define the degrees of freedom (df) as the number of effects used to find the sum of squares. This is 5 in this case.

The standard error is the square root of the mean squares of the effects. The formula for finding the standard error is

$$\text{Standard error} = \sqrt{\frac{SS}{df}}$$

We find here that standard error $= \sqrt{0.375/5} = 0.273\,86$. We can then find a Student t variable with df degrees of freedom.

To choose an example, we test the hypothesis that the effect AB is 0:

$$t_5 = \frac{AB - 0}{\text{Standard error}}$$
$$t_5 = \frac{0.5000 - 0}{0.27386}$$
$$t_5 = 1.8258$$

The p-value for the test is the probability that this value of t is exceeded or that t is at most -1.8258. This is 0.127464, so we probably would not decide that this is a statistically significant effect.

It is probably best to use a computer program or a statistical calculator to determine the p-values since only crude estimates of the p-values can be made using tables. We used the computer algebra program *Mathematica* to determine the t values and the corresponding p-values in Table 15.9.

Table 15.9

Effect	Size	t_5	p-Value
μ	72.25	263.821	$1.48 \cdot 10^{-11}$
A	-4	-14.606	$2.72 \cdot 10^{-5}$
B	12	43.818	$1.17 \cdot 10^{-7}$
AB	0.50	1.826	0.127
C	-1.125	-4.108	0.009
AC	0.375	1.369	0.229
BC	-0.625	-2.282	0.071
D	-2.75	-10.042	$1.68 \cdot 10^{-4}$
AD	0	0	1.00
BD	2.25	8.216	$4.34 \cdot 10^{-4}$
CD	-0.125	-0.456	0.667

Using the critical value of 0.05 as our level of significance, we would conclude that μ, A, B, C, D, and BD are of statistical significance while the other interactions can be safely ignored. This may have important consequences for the experimenter as future experiments are planned.

Statistical computer programs are of great value in analyzing experimental designs. Their use is almost mandatory when the number of main factors is 5 or more. These programs also provide graphs that can give great insight into the data. For example, the statistical computer program Minitab generates a graph of the main effects and interactions for the example we have been considering, which is shown in Figure 15.6.

Significant effects are those that are not close to the straight line shown. While the mean is not shown, we would draw the same conclusions from the graph as we did above.

Most experiments have more than one observation for each factor level combination. In that case, the determination of the statistical significance of the factors and interactions is much more complex than in the case we have discussed. This is a topic commonly considered in more advanced courses in statistics. Such courses might also consider other types of experimental designs that occur in science and engineering.

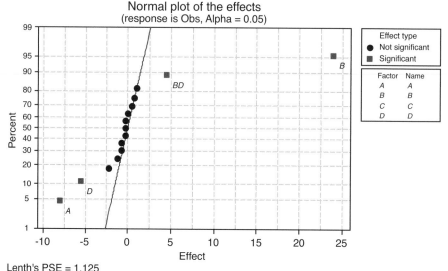

Lenth's PSE = 1.125

Figure 15.6

CONCLUSIONS

This has been a very brief introduction to the design of experiments. Much more is known about this subject and the interested reader is referred to more advanced books on this subject. We have made use of the geometry here in analyzing experimental data since that provides a visual display of the data and the conclusions we can draw from the data.

EXPLORATIONS

1. Check that the model given after Table 15.4 predicts each of the observations exactly.
2. Show that collapsing the cubes shown in Figure 15.3 to squares leads to the determination of the two-factor interactions.
3. In Example 15.4, confound using $ABCD = -1$ and analyze the resulting data.
4. The following data represent measurements of the diameter of a product produced at two different plants and on three different shifts.

	Shift		
	1	2	3
Plant 1	66.2	66.3	64.2
	64,7	65.7	64.7
Plant 2	65.3	65.4	66.2
	61.5	63.4	67.2

Analyze the data and state any conclusions that can be drawn. Find all the main effects and the interactions and show how the data can be predicted using a linear model.

Chapter 16

Recursions and Probability

CHAPTER OBJECTIVES:

- to learn about recursive functions
- to apply recursive functions to permutations and combinations
- to use recursive functions to find the expected value of the binomial distribution
- to learn how to gamble (perhaps wisely)
- consider the occurrence of *HH* in coin tossing.

INTRODUCTION

Since they are very useful in probability, we consider functions whose values depend upon other values of the same function. Such functions are called *recursive*. These functions are investigated here in general, and then we show their application to probability and probability distribution functions.

EXAMPLE 16.1   *A General Recursion*

We begin with a nonprobability example. Suppose we define a function on the positive integers, $f(x)$, where

$$f(x+1) = 2f(x), \quad x = 0, 1, 2, \ldots$$

If we have a starting place, say $f(1) = 1$, then we can determine any of the subsequent values for the function. For example,

$$f(2) = 2f(1) = 2 \cdot 1 = 2, \; f(3) = 2f(2) = 2 \cdot 2 = 4, \; f(4) = 2 \cdot f(3) = 2 \cdot 4 = 8$$

and so on. It is easy to see here, since subsequent values of f are twice that of the preceding value, that the values of f are powers of 2, and that in fact

$$f(x) = 2^{x-1} \quad \text{for } x = 2, 3, 4, \ldots$$

A Probability and Statistics Companion, John J. Kinney
Copyright © 2009 by John Wiley & Sons, Inc.

The relationship $f(x+1) = 2f(x)$ is called a *recursion* or *difference equation*, and the formula $f(x) = 2^{x-1}$ is its solution. The solution was probably evident from the start.

To verify that $f(x) = 2^{x-1}$ for $x = 2, 3, 4, \ldots$ is the solution, note that if $f(x) = 2^{x-1}$, then $f(x+1) = 2^x = 2 \cdot 2^{x-1} = 2f(x)$.

Finding solutions for recursions, however, is not always so easy. Consider

$$f(x+1) = 2f(x) + f(x-1) \quad \text{for } x = 1, 2, 3 \ldots$$

The solution is far from evident. Let us start by finding some of the values for f, but this time we need two starting points, say $f(0) = 1$ and $f(1) = 2$. Then, applying the recursion, we find

$$f(2) = 2f(1) + f(0) = 2 \cdot 2 + 1 = 5$$

$$f(3) = 2f(2) + f(1) = 2 \cdot 5 + 2 = 12$$

$$f(4) = 2f(3) + f(2) = 2 \cdot 12 + 5 = 29,$$

and so on. ∎

It is easy to write a computer program to produce a table of these results.

Analytic methods exist to produce solutions for the recursions we consider here, and while we will not explore them, we invite the reader to check that the solutions we present are, in fact, solutions of the recursions. We did this in our first example.

In this case, the solution of the recursion is

$$f(x) = \frac{(1+\sqrt{2})^{x+1} - (1-\sqrt{2})^{x+1}}{2\sqrt{2}}, \quad x = 0, 1, 2, \ldots$$

Those $\sqrt{2}$'s look troublesome at first glance, but they all disappear! Table 16.1 shows some of the values of $f(x)$ obtained from this solution.

Table 16.1

x	$f(x)$
0	1
1	2
2	5
3	12
4	29
5	70
6	169
7	408
8	985
9	2378
10	5741

The values now increase rather rapidly. A computer program finds that

$$f(100) = 161, 733, 217, 200, 188, 571, 081, 311, 986, 634, 082, 331, 709$$

It is difficult to think about computing this in any other way. Now we turn to probability, the subject of all our remaining applications.

EXAMPLE 16.2   *Permutations*

An easy application of recursions is with *permutations*—an arrangement of objects in a row. If we have n distinct items to arrange, suppose we denote the number of distinct permutations by $_n P_n$. Since we know there are $n!$ possible permutations, it follows that $_n P_n = n!$

It is also clear that

$$n! = n \cdot (n-1)!$$

so

$$_n P_n = n \cdot_{n-1} P_{n-1}, \text{ a recursion}$$

Since we know that $1! = 1$, we can use the above result to find that and subsequent values. Ordinarily, in finding $4!$ we would have to multiply $4 \cdot 3 \cdot 2 \cdot 1$, and in finding $5!$ we would have to calculate $5 \cdot 4 \cdot 3 \cdot 2 \cdot 1$. It is much easier and faster to use the facts that $4! = 4 \cdot 3!$ and $5! = 5 \cdot 4!$ since it is easy to calculate $3!$ and $4!$. ■

It is easy to continue and produce a table of factorials very quickly and easily. We continue now and show more examples of recursions and their uses in probability.

EXAMPLE 16.3   *Combinations*

A *combination* is the number of distinct samples, say of size r, that can be selected from a set of n distinct items. This number is denoted by $\binom{n}{r}$. We know that

$$\binom{n}{r} = \frac{n!}{r! \cdot (n-r)!} \quad \text{where } 0 \le r \le n$$

To find a recursion, note that

$$\binom{n}{r+1} = \frac{n!}{(r+1)! \cdot (n-r-1)!}$$

so

$$\frac{\binom{n}{r+1}}{\binom{n}{r}} = \frac{n!}{(r+1)! \cdot (n-r-1)!} \cdot \frac{r!(n-r)!}{n!} = \frac{n-r}{r+1}$$

which we can write in recursive form as

$$\binom{n}{r+1} = \frac{n-r}{r+1} \cdot \binom{n}{r}$$

■

Using this recursion will allow us to avoid calculation of any factorials! To see this, note that $\binom{n}{1} = n$. So it follows that

$$\binom{n}{2} = \frac{n-1}{1+1} \cdot \binom{n}{1}$$

or

$$\binom{n}{2} = \frac{n-1}{2} \cdot n = \frac{n(n-1)}{2}$$

We can continue this by calculating

$$\binom{n}{3} = \frac{n-2}{2+1} \cdot \binom{n}{2}$$

or

$$\binom{n}{3} = \frac{n-2}{3} \cdot \frac{n(n-1)}{2} = \frac{n(n-1)(n-2)}{3 \cdot 2}$$

and

$$\binom{n}{4} = \frac{n-3}{3+1} \cdot \binom{n}{3}$$

or

$$\binom{n}{4} = \frac{n-3}{4} \cdot \frac{n(n-1)(n-2)}{3 \cdot 2} = \frac{n(n-1)(n-2) \cdot (n-3)}{4 \cdot 3 \cdot 2}$$

To take a specific example, suppose $n = 5$. Since $\binom{5}{1} = 5$, the recursion can be used repeatedly to find

$$\binom{5}{2} = \frac{5-1}{1+1} \cdot \binom{5}{1}$$

or

$$\binom{5}{2} = \frac{4}{2} \cdot 5 = 10$$

and

$$\binom{5}{3} = \frac{5-2}{2+1} \cdot \binom{5}{2}$$

or

$$\binom{5}{3} = \frac{3}{3} \cdot 10 = 10$$

and we could continue this. When the numbers become large, it is a distinct advantage not to have to calculate the large factorials involved. We continue now with some probability distribution functions.

EXAMPLE 16.4   *Binomial Probability Distribution Function*

It is frequently the case in probability that one value of a probability distribution function can be found using some other value of the probability distribution function. Consider the binomial probability distribution function as an example.

For the binomial probability distribution function, we know

$$f(x) = P(X = x) = \binom{n}{x} p^x q^{n-x}, \quad x = 0, 1, \ldots, n$$

Now

$$P(X = x + 1) = \binom{n}{x+1} p^{x+1} q^{n-(x+1)}$$

so if we divide $P(X = x + 1)$ by $P(X = x)$, we find

$$\frac{P(X = x + 1)}{P(X = x)} = \frac{\binom{n}{x+1} p^{x+1} q^{n-(x+1)}}{\binom{n}{x} p^x q^{n-x}} = \frac{n!}{(x+1)!(n-x-1)!} \cdot \frac{x!(n-x)!}{n!} \cdot \frac{p}{q}$$

which simplifies to

$$\frac{P(X = x + 1)}{P(X = x)} = \frac{n-x}{x+1} \cdot \frac{p}{q}$$

and this can be written as

$$P(X = x + 1) = \frac{n-x}{x+1} \cdot \frac{p}{q} \cdot P(X = x)$$

The recursion is very useful; for example, we know that $P(X = 0) = q^n$. The recursion, using $x = 0$, then tells us that

$$P(X = 1) = \frac{n-0}{0+1} \cdot \frac{p}{q} \cdot P(X = 0)$$

$$P(X = 1) = n \cdot \frac{p}{q} \cdot q^n = n \cdot p \cdot q^{n-1}$$

which is the correct result for $X = 1$.

This can be continued to find $P(X = 2)$. The recursion tells us that

$$P(X = 2) = \frac{n-1}{1+1} \cdot \frac{p}{q} \cdot P(X = 1)$$

$$P(X = 2) = \frac{n-1}{2} \cdot \frac{p}{q} \cdot n \cdot p \cdot q^{n-1}$$

$$P(X = 2) = \frac{n(n-1)}{2} \cdot p^2 \cdot q^{n-2}$$

$$P(X = 2) = \binom{n}{2} \cdot p^2 \cdot q^{n-2}$$

and again this is the correct result for $P(X = 2)$.   ■

We can continue in this way, creating all the values of the probability distribution function.

The advantage in doing this is that the quantities occurring in the values of the probability distribution function do not need to be calculated each time. The value, for example, of $n!$ is never calculated at all. To take a specific example, suppose that $p = 0.6$ so that $q = 0.4$ and that $n = 12$.

Letting X denote the number of successes in 12 trials, we start with

$$P(X = 0) = 0.4^{12} = 0.000016777$$

The recursion is

$$P(X = x+1) = \frac{n-x}{x+1} \cdot \frac{p}{q} \cdot P(X = x)$$

which in this case is

$$P(X = x+1) = \frac{12-x}{x+1} \cdot \frac{0.6}{0.4} \cdot P(X = x)$$

so

$$P(X = 1) = 12 * 1.5 * 0.000016777 = 0.00030199$$

and

$$P(X = 2) = \frac{11}{2} * 1.5 * 0.00030199 = 0.0024914$$

We can continue and find the complete probability distribution function in Table 16.2.

Table 16.2

x	$P(X = x)$
0	0.000017
1	0.000302
2	0.002491
3	0.012457
4	0.042043
5	0.100902
6	0.176579
7	0.227030
8	0.212841
9	0.141894
10	0.063852
11	0.017414
12	0.002177

EXAMPLE 16.5   *Finding the Mean of the Binomial*

We actually used a recursion previously in finding the mean and the variance of the negative hypergeometric distribution in Chapter 7.

One interesting application of the recursion

$$P(X = x + 1) = \frac{n - x}{x + 1} \cdot \frac{p}{q} \cdot P(X = x)$$

lies in finding the mean of the binomial probability distribution. Rearranging the recursion and summing the recursion from 0 to $n - 1$ gives

$$\sum_{x=0}^{n-1} q(x + 1)P(X = x + 1) = \sum_{x=0}^{n-1} p(n - x)P(X = x)$$

which can be written and simplified as

$$qE[X] = np \sum_{x=0}^{n-1} P(X = x) - p \sum_{x=0}^{n-1} x \cdot P(X = x)$$

which becomes

$$qE[X] = np[1 - P(X = n)] - p[E[X] - nP(X = n)]$$

or

$$qE[X] = np - npq^n - pE[X] + npq^n$$

and this simplifies easily to

$$E[X] = np$$

■

This derivation is probably no simpler than the standard derivation that evaluates

$$E[X] = \sum_{x=0}^{n} x \cdot P(X = x),$$

but it is shown here since it can be used with any discrete probability distribution, usually providing a derivation easier than the direct calculation.

EXAMPLE 16.6   *Hypergeometric Probability Distribution Function*

We show one more discrete probability distribution and a recursion.

Suppose that a lot of N manufactured products contains D items that are special in some way. The sample is of size n. Let the random variable X denote the number of special items in the sample that is selected with nonreplacement, that is, a sampled item is not returned to the lot before the next item is drawn. The probability distribution function is

$$P(X = x) = \frac{\binom{D}{x}\binom{N - D}{n - x}}{\binom{N}{n}}, \quad x = 0, 1, 2, \ldots, \, Min[n, D]$$

A recursion is easily found since by considering $P(X = x)/P(X = x - 1)$ and simplifying, we find

$$P(X = x) = \frac{(D - x + 1)(n - x + 1)}{x(N - D - n + x)} P(X = x - 1)$$

To choose a specific example, suppose $N = 100$, $D = 20$, and $n = 10$.
Then

$$P(X = 0) = \frac{(N - D)!(N - n)!}{(N - D - n)!N!} = \frac{80!90!}{70!100!} = \frac{80 \cdot 79 \cdot 78 \cdot \cdots \cdot 71}{100 \cdot 99 \cdot 98 \cdot \cdots \cdot 91} = 0.0951163$$

Applying the recursion,

$$P(X = 1) = \frac{D \cdot n}{(N - D - n + 1)} P(X = 0) = \frac{20 \cdot 10}{71} \cdot 0.0951163 = 0.267933$$

and

$$P(X = 2) = \frac{(D - 1) \cdot (n - 1)}{2(N - D - n + 2)} P(X = 1) = \frac{19 \cdot 9}{2 \cdot 72} \cdot 0.267933 = 0.318171$$

∎

This could be continued to give all the values of the probability distribution function. Note that although the definition of $P(X = x)$ involves combinations, and hence factorials, we never computed a factorial!

The recursion can also be used, in a way entirely similar to that we used with the binomial distribution, to find that the mean value is $E[X] = (n \cdot D)/N$. Variances can also be produced this way.

EXAMPLE 16.7 *Gambler's Ruin*

We now turn to an interesting probability situation, that of the Gambler's Ruin.

Two players, A and B, play a gambling game until one player is out of money; that is, the player is ruined. Suppose that $1 is gained or lost at each play, and A starts with a , B starts with b, and $a + b = N$. Let p_k be the probability that A (or B) is ruined with a fortune of k.

Suppose that the player has k and the probability of winning $1 on any given play is p and the probability of losing on any given play is then $1 - p = q$. If the player wins on the next play, then his or her fortune is $(k + 1)$, while if the next play produces a loss, then his or her fortune is $(k - 1)$. So

$$p_k = pp_{k+1} + qp_{k-1}$$

where $p_0 = 1$ and $p_N = 0$

The solution of this recursion depends on whether p and q are equal or not,

If $p \neq q$ then the solution is

$$p_k = \frac{\left(\dfrac{q}{p}\right)^k - \left(\dfrac{q}{p}\right)^N}{1 - \left(\dfrac{q}{p}\right)^N} \quad \text{for } k = 0, 1, 2, \ldots, N$$

If $p = q$ then the solution is

$$p_k = 1 - \frac{k}{N} \quad \text{for } k = 0, 1, 2, \ldots, N$$

Let us consider the fair game (where $p = q = 1/2$) and player A whose initial fortune is a. The probability the player is ruined is then

$$p_a = 1 - \frac{a}{N} = 1 - \frac{a}{a+b} = \frac{b}{a+b} = \frac{1}{\dfrac{a}{b} + 1}$$

This means that if A is playing against a player with a relatively large fortune (so $b \gg a$), then $p_a \longrightarrow 1$ and the player faces almost certain ruin. The only question is how long the player will last. This is the case in casino gambling where the house's fortune is much greater than that of the player. Note that this is for playing a fair game, which most casino games are not.

Now let us look at the case where $p \neq q$, where the solution is

$$p_k = \frac{\left(\dfrac{q}{p}\right)^k - \left(\dfrac{q}{p}\right)^N}{1 - \left(\dfrac{q}{p}\right)^N} \quad \text{for } k = 0, 1, 2, \ldots, N$$

The probability of ruin now depends on two ratios—that of q/p as well as a/b. Individual calculations are not difficult and some results are given in Table 16.3. Here, we have selected $q = 15/28$ and $p = 13/28$, so the game is slightly unfair to A.

So the probability of ruin is quite high almost without regard for the relative fortunes. A graph of the situation is also useful and shown in Figure 16.1, where p denotes the probability that favored player, here B, wins. ∎

Table 16.3

$a	$b	p_a
15	10	0.782808
15	20	0.949188
20	10	0.771473
20	20	0.945937
25	20	0.944355
25	25	0.972815
30	25	0.972426
30	30	0.986521
35	30	0.986427
35	35	0.993364
40	35	0.993341
40	40	0.996744

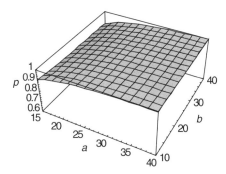

Figure 16.1

EXAMPLE 16.8   *Finding a Pattern in Binomial Trials*

It is interesting to explore patterns when binomial trials are performed. A perfect model for this is tossing a coin, possibly a loaded one, and looking at the pattern of heads and tails. We previously considered waiting for the first occurrence of the pattern *HH* in Chapter 7. In this example, we look only for the occurrence of the pattern *HH*, not necessarily the first occurrence. First, we need to define when this pattern "occurs". Consider the sequence *TTHHHTTTHHHHHH*. Scan the sequence from left to right and we see that *HH* occurs on the 4th flip. Then begin the sequence all over again. The pattern occurs again on the 10th, 12th, and 14th trials and at no other trials.

Let u_n denote the probability the sequence *HH* occurs at the nth trial. If a sequence ends in HH, then either the pattern occurs at the nth trial or it occurs at the $(n-1)$st trial and is followed by H. So, since all possible sequences ending in *HH* have probability p^2, we have the recursion

$$u_n + pu_{n-1} = p^2 \quad \text{for } n \geq 2, \quad u_1 = 0$$

The reader can check that the solution to this recursion is

$$u_n = \frac{p}{1+p}[p+(-p)^n] \quad \text{for } n \geq 2$$

Some of the values this gives are

$$u_2 = p^2; \quad u_3 = qp^2; \quad u_4 = p^2(p^2 - p + 1); \quad u_5 = \frac{p}{1+p}\left[p - p^5\right];$$

$$u_6 = \frac{p}{1+p}\left[p + p^6\right]; \quad u_7 = \frac{p}{1+p}\left[p - p^7\right]$$

and so on, a really beautiful pattern.

Since $(-p)^n$ becomes small as n becomes large, it is evident that $u_n \; -> \; p^2/(1+p)$. If $p = 1/2$, then $u_n \; -> \; 1/6$. This limit occurs fairly rapidly as the values in Table 16.4 show. ∎

Table 16.4

n	u_n
2	0.25
4	0.1875
6	0.171875
8	0.167969
10	0.166992
12	0.166748
14	0.166687
16	0.166672
18	0.166668
20	0.166667

CONCLUSIONS

Recursions, or difference equations, are very useful in probability and can often be used to model situations in which the formation of probability functions is challenging. As we have seen, the recursions can usually be solved and then calculations made.

EXPLORATIONS

1. Verify that the solution for $f(x + 12f(x) + f(x - 1)$, $f(0) = 1$, $f(1) = 2$ is that given in the text.

2. Show how to use the recursion $\binom{n}{r+1} = \left[(n - r)/(r - 1)\right]\binom{n}{r}$.

3. Establish a recursion for $\binom{n+1}{r}/\binom{n}{r}$ and show an application of the result.

4. Use Example 16.6 to find a recursion for the hypergeometric distribution and use it to find its mean value.

5. The *Poisson* distribution is defined as $f(x) = e^{-\mu}\mu^x/x!$ for $x = 0, 1, 2, \cdots$ and $\mu > 0$. Find the recursion for the values of $f(x)$ and use it to establish the fact that $E(X) = \mu$.

Chapter 17

Generating Functions and the Central Limit Theorem

CHAPTER OBJECTIVES:

- to develop the idea of a *generating function* here and show how these functions can be used in probability modeling
- to use generating functions to investigate the behavior of sums
- to see the development of the central limit theorem.

EXAMPLE 17.1 *Throwing a Fair Die*

Let us suppose that we throw a fair die once. Consider the function $g(t) = (1/6)t + (1/6)t^2 + (1/6)t^3 + (1/6)t^4 + (1/6)t^5 + (1/6)t^6$ and the random variable X denoting the face that appears.

The coefficient of t^k in $g(t)$ is $P(X = k)$ for $k = 1, 2, \ldots, 6$. This is shown in Figure 17.1.

Consider tossing the die twice with X_1 and X_2, the relevant random variables. Now calculate $[g(t)]^2$. This is

$$[g(t)]^2 = \frac{1}{36}[t^2 + 2t^3 + 3t^4 + 4t^5 + 5t^6 + 6t^7 + 5t^8 + 4t^9 + 3t^{10} + 2t^{11} + t^{12}]$$

The coefficient of t^k is now $P(X_1 + X_2 = k)$. A graph of this is interesting and is shown in Figure 17.2.

This process can be continued, the coefficients of $[g(t)]^3$ giving the probabilities associated with the sum when three fair dice are thrown. The result is shown in Figure 17.3.

This "normal-like" pattern continues. Figure 17.4 shows the sums when 24 fair dice are thrown.

Since the coefficients represent probabilities, $g[t]$ and its powers are called *generating functions*. ∎

A Probability and Statistics Companion, John J. Kinney
Copyright © 2009 by John Wiley & Sons, Inc.

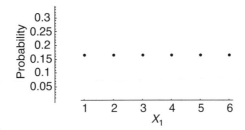

Figure 17.1

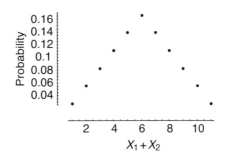

Figure 17.2

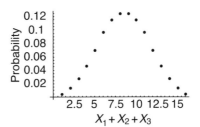

Figure 17.3

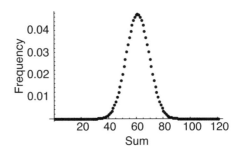

Figure 17.4

MEANS AND VARIANCES

This behavior of sums is a consequence of the central limit theorem, which states that the probability distribution of sums of independent random variables approaches a normal distribution. If these summands, say $X_1, X_2, X_3, \ldots, X_n$, have means $\mu_1, \mu_2, \mu_3, \ldots, \mu_n$ and variances $\sigma_1^2, \sigma_2^2, \sigma_3^2, \ldots, \sigma_n^2$, then $X_1 + X_2 + X_3 + \cdots + X_n$ has expectation $\mu_1 + \mu_2 + \mu_3 + \cdots + \mu_n$ and variance $\sigma_1^2 + \sigma_2^2 + \sigma_3^2 + \cdots + \sigma_n^2$.

Our example illustrates this nicely. Each of the $X_i's$ has the same uniform distribution with $\mu_i = 7/2$ and $\sigma_i^2 = 35/12$, $i = 1, 2, 3, \ldots, n$.

So we find the following means and variances in Table 17.1 for various numbers of summands.

Table 17.1

n	μ	σ^2
1	7/2	35/12
2	7	35/6
3	21/2	35/4
24	84	70

EXAMPLE 17.2   *Throwing a Loaded Die*

The summands in the central limit theorem need not all have the same mean or variance. Suppose the die is loaded and the generating function is

$$h(t) = \frac{t}{10} + \frac{t^2}{5} + \frac{t^3}{20} + \frac{t^4}{20} + \frac{t^5}{5} + \frac{2t^6}{5}$$

A graph of this probability distribution, with variable X again, is shown is Figure 17.5.

When we look at sums now, the normal-like behavior does not appear quite so soon. Figure 17.6 shows the sum of three of the loaded dice.

But the normality does appear. Figure 17.7 shows the sum of 24 dice.

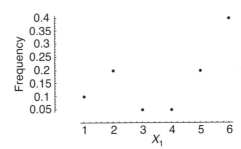

Figure 17.5

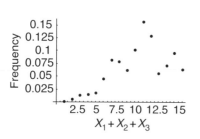

Figure 17.6

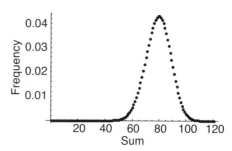

Figure 17.7

The pattern for the mean and variances of the sums continues.

$$\mu_1 = \frac{17}{4} \text{ and } \mu_{24} = 102 \quad \text{while } \sigma_1^2 = \frac{279}{80} \text{ and } \sigma_{24}^2 = \frac{837}{10}$$

■

A NORMAL APPROXIMATION

Let us return now to the case of the fair die and the graph of the sum of 24 tosses of the die as shown in Figure 17.4.

Since $E[X_1] = 7/2$ and $\text{Var}[X_1] = 35/12$, we know that

$$E[X_1 + X_2 + X_3 + \cdots + X_{24}] = 24 \cdot \frac{7}{2} = 84$$

and

$$\text{Var}[X_1 + X_2 + X_3 + \cdots + X_{24}] = 24 \cdot \frac{35}{12} = 70$$

The visual evidence in Figure 17.4 suggests that the distribution becomes very normal-like so we compare the probabilities when 24 dice are thrown with values of a normal distribution with mean 84 and standard deviation $\sqrt{70} = 8.3666$. A comparison is shown in Table 17.2. S denotes the sum.

So the normal approximation is excellent.

This is yet another illustration of the central limit theorem. Normality, in fact, occurs whenever a result can be considered as the sum of independent random variables. We find that many human characteristics such as height, weight, and IQ are normally distributed. If the measurement can be considered to be the result of the sum of factors, then the central limit theorem assures us that the result will be normal.

Table 17.2

S	Probability	Normal
72	0.0172423	0.0170474
73	0.0202872	0.0200912
74	0.0235250	0.0233427
75	0.0268886	0.0267356
76	0.0302958	0.0301874
77	0.0336519	0.0336014
78	0.0368540	0.036871
79	0.0397959	0.0398849
80	0.0423735	0.0425331
81	0.0444911	0.0447139
82	0.0460669	0.0463396
83	0.0470386	0.0473433
84	0.047367	0.0476827
85	0.0470386	0.0473433
86	0.0460669	0.0463396
87	0.0444911	0.0447139
88	0.0423735	0.0425331
89	0.0397959	0.0398849
90	0.0368540	0.036871
91	0.0336519	0.0336014
92	0.0302958	0.0301874
93	0.0268886	0.0267356
94	0.0235250	0.0233427
95	0.0202872	0.0200912

CONCLUSIONS

Probability generating functions are very useful in calculating otherwise formidable probabilities. They are also helpful in indicating limiting probability distributions. We have shown here primarily the occurrence of the central limit theorem in finding the limiting normal distributions when sums of independent random variables are considered.

EXPLORATIONS

1. Load a die in any way you would like so long as it is not fair, investigate the sums when the die is tossed a few times, and then investigate the behavior of the die when it is tossed a large number of times.
2. Find the mean and variance of the loaded die in Exploration 1 and verify the mean and variance of the sums found in Exploration 1.

3. Show that the normal curve is a very good approximation for the sums on your loaded die when tossed a large number of times.

4. Show that the function $p(t) = (q + pt)^n$ generates probabilities for the binomial random variable with parameters q, p, and n.

5. Use the generating function in Exploration 4 to find the mean and variance of the binomial random variable.

6. The geometric random variable describes the waiting time until the first success in a binomial situation with parameters p and q. Its probability distribution function is $f(x) = pq^{x-1}$, $x = 1, 2, \ldots$ Show that its probability generating function is $pt/(1 - qt)$ and use this to find the mean and the variance of the geometric random variable.

Bibliography

WHERE TO LEARN MORE

There is now a vast literature on the theory of probability. A few of the following references are cited in the text; other titles that may be useful to the instructor or student are included here as well.

1. G. E. P. Box, W. G. Hunter, and J. Stuart Hunter, *Statistics for Experimenters*, John Wiley & Sons, 1978.
2. F. N. David and D. E. Barton, *Combinatorial Chance*, Charles Griffin & Company Limited, 1962.
3. J. W. Drane, S. Cao, L. Wang, and T. Postelnicu, Limiting forms of probability mass functions via recurrence formulas, *The American Statistician*, 1993, 47(4), 269–274.
4. N.R. Draper and H. Smith, *Applied Regression Analysis*, second edition, John Wiley & Sons, 1981.
5. A. J. Duncan, *Quality Control and Industrial Statistics*, fifth edition, Richard D. Irwin, Inc., 1896.
6. W. Feller, *An Introduction to Probability and Its Applications*, Volumes I and II, John Wiley & Sons, 1968.
7. B. V. Gnedenko, *The Theory of Probability*, fifth edition, Chelsea Publishing Company, 1989.
8. S. Goldberg, *Probability: An Introduction*, Prentice-Hall, Inc., 1960.
9. S. Goldberg, *Introduction to Difference Equations*, Dover Publications, 1986.
10. E. L. Grant and R. S. Leavenworth, *Statistical Quality Control*, sixth edition, McGraw-Hill, 1988.
11. R. P. Grimaldi, *Discrete and Combinatorial Mathematics*, fifth edition, Addison-Wesley Publishing Co., Inc., 2004.
12. A. Hald, *A History of Probability and Statistics and Their Applications Before 1750*, John Wiley & Sons, 1990.
13. N. L. Johnson, S. Kotz, and A. W. Kemp, *Univariate Discrete Distributions*, second edition, John Wiley & Sons, 1992.
14. N. L. Johnson, S. Kotz, and N. Balakrishnan, *Continuous Univariate Distributions*, Volumes 1 and 2, second edition, John Wiley & Sons, 1994.
15. J. J. Kinney, Tossing coins until all are heads, *Mathematics Magazine*, May 1978, pp. 184–186.
16. J. J. Kinney, *Probability: An Introduction with Statistical Applications*, John Wiley & Sons, 1997.
17. J. J. Kinney, *Statistics for Science and Engineering*, Addison-Wesley Publishing Co., Inc., 2002.
18. F. Mosteller, *Fifty Challenging Problems in Probability*, Addison-Wesley Publishing Co., Inc., 1965. Reprinted by Dover Publications.
19. S. Ross, *A First Course in Probability*, sixth edition, Prentice-Hall, Inc., 2002.

A Probability and Statistics Companion, John J. Kinney
Copyright © 2009 by John Wiley & Sons, Inc.

20. D. Salsburg, *The Lady Tasting Tea: How Science Revolutionized Science in the Twentieth Century*, W. H. Freeman and Company, 2001.
21. J. V. Uspensky, *Introduction to Mathematical Probability*, McGraw-Hill Book Company, Inc., 1937.
22. W. A. Whitworth, *Choice and Chance*, fifth edition, Hafner Publishing Company, 1965.
23. S. Wolfram, *Mathematica: A System for Doing Mathematics by Computer*, Addison-Wesley Publishing Co., Inc., 1991.

Index

Acceptance sampling   32
Addition Theorem   10, 25
α 135
Alternative hypothesis   134
Analysis of variance   197
Average outgoing quality   34
Axioms of Probability   8

Bayes' theorem   45
β 137
Binomial coefficients   24
Binomial probability
    distributuion   64, 244
  Mean   69
  Variance   69
Binomial theorem   3, 11, 15, 24, 83
Birthday problem   8, 16
Bivariate random
      variables   115

Cancer test   42
Central limit theorem   121, 123, 213
Cereal Box Problem   88
Chi squared distribution   141
Choosing an assistant   30
Combinations   22, 242
Conditional probability   10, 39, 40
Confidence coefficient   131, 149
Confidence interval   131
Confounding   233
Control charts   155
  Mean   159, 160
  np 161
  p 163
Correlation coefficient   200
Counting principle   19
Critical region   135

Degrees of freedom   143
Derangements   17
Difference equation   241
Discrete uniform distriution   59
Drivers'ed   39

e   5, 6, 26, 31
Estimation   130, 177
Estimating σ 157, 159
Expectation   6
  Binomial distribution   69
  Geometric distribution   73
  Hypergeometric distribution   71
  Negative Binomial distribution   88
  Negative Hypergeometric
    distribution   102

F ratio distribution   148
Factor   224
Factorial experiment   231, 232
Finite population correction
    factor   71, 214
Fractional factorial experiment   234

Ganbler's ruin   248
General addition theorem   10, 25
Generating functions   252
Geometric probability   48
Geometric probability distribution   72, 84
Geometric series   12, 13
German tanks   28, 62, 177

Hat problem   5, 26
Heads before tails   88
Hypergeometric probability distribution
    70, 105, 247
Hypothesis testing   133

Inclusion and exclusion   26
Independent events   11
Indianapolis 500 data   196
Influential observations   193
Interaction   225

Joint probability distributions   115

Least squares   191
Let's Make a Deal   8, 15, 17, 44
Lower control limit   158
Lunch problem   96

Marginal distributions   117
Maximum   176
Mean   60
  Binomial distribution   69, 246
  Geometric distribution   73
  Hypergeometric distribution   71
  Negative Binomial distribution   88
  Negative Hypergeometric
    distribution   174
Mean of means   124
Means - two samples   150
Median   28, 174
Median - median line   202, 207
Meeting at the library   48
Minimum   174, 177
Multiple regression   235
Multiplication rule   10
Mutually exclusive events   9
Mythical island   84

Negative binomial distribution   87, 103
Negative hypergeometric distribution   99
Nonlinear models   201
Nonparametric methods   170
Normal probability distribution   113
Null hypothesis   134

Operating characteristic curve   138
Optimal allocation   217
Order statistics   173, 174

p-value for a test   139
Patterns in binomial trials   90
Permutations   5, 12, 19, 242
  Some objects alike   20
Poisson distribution   250
Poker   27

Pooled variances   152
Power of a test   137
Probability   2, 8
Probability distribution   32, 59
Proportional allocation   215

Race cars   28, 64
Random variable   58
Randomized response   46, 55
Range   156, 176
Rank sum test   170
Ratio of variances   148
Recursion   91, 101, 241
Residuals   189
Runs   3, 180, 182
  Theory   182
  Mean   184
  Variance   184

Sample space   2, 4, 6, 7, 10, 14
Sampling   211
Seven game series in sports   75
Significance level   135
Simple random sampling   211
Standard deviation   60
Strata   214, 221
Student t distribution   146
Sums   62, 69, 111, 121

Type I error   135
Type II error   135

Unbiased estimator   212
Uniform random variable   59, 109
Upper control limit   158

Variance   60, 119
  Binomial distribution   69
  Hypergeometric distribution   71
  Negative Binomial distribution   88
  Negative Hypergeometric
    distribution   102

Waiting at the library   48
Waiting time problems   83
World series   76

Yates algorithm   230